S0-ABS-089

Effective Strategic Planning for Competitive Advantage

Ten Steps for Technical Professions

DAVID L. GOETSCH

CEO, Institute for Continual Improvement

NETEFFECT SERIES

PEARSON

Prentice
Hall

Upper Saddle River, New Jersey
Columbus, Ohio

Library of Congress Cataloging-in-Publication Data

Goetsch, David L.

  Effective strategic planning for competitive advantage : ten steps for technical professions / by David L. Goetsch.

  p. cm.

  ISBN 0-13-048525-X

  1. Strategic planning. 2. Competition. I. Title.

HD30.28.G64 2006

658.4'012--dc22   2004024424

Executive Editor: Debbie Yarnell
Production Editor: Louise N. Sette
Production Supervision: Carlisle Publishers Services
Design Coordinator: Diane Ernsberger
Cover Designer: Keith Van Norman
Production Manager: Deidra M. Schwartz
Marketing Manager: Jimmy Stephens

This book was set in GoudyWtcTReg by Carlisle Communications, Ltd. It was printed and bound by Phoenix Color Book Group. The cover was printed by Phoenix Color Book Group.

Pearson Education Ltd.
Pearson Education Singapore Pte. Ltd.
Pearson Education Canada, Ltd.
Pearson Education—Japan

Pearson Education Australia Pty. Limited
Pearson Education North Asia Ltd.
Pearson Educación de Mexico, S.A. de C.V.
Pearson Education Malaysia Pte. Ltd.

10 9 8 7 6 5 4 3 2 1
ISBN 0-13-048525-X

Contents

About the Author

David L. Goetsch is President and CEO of the Institute for Continual Improvement, a private consulting firm dedicated to the continual improvement of employees, organizations, and communities. Dr. Goetsch welcomes feedback from his readers and may be reached at the following e-mail address: ddsg2001@cox.net.

Introduction
Effective Strategic Planning—A Ten-Step Model

To survive and thrive in today's intensely competitive world of business, organizations must establish and sustain competitive advantage in the marketplace. The process for establishing competitive advantage is known as *strategic planning*. Competitive advantage is established in the conceptual sense during the strategic planning process. But conceptualizing competitive advantage and actually achieving it are two very different propositions. To actually achieve and sustain competitive advantage, organizations must *effectively execute* their strategic plans.

The strategic planning process answers three basic, but critical, questions for organizations:

1. Who are we?
2. Where are we going?
3. How will we get there?

To answer these questions, an organization must conduct a thorough internal self-assessment that reveals strengths, weaknesses, financial condition, and core competencies that produce value. It is also necessary to conduct a comprehensive external assessment of competitors and to make informed predictions about future market behavior. Using the information collected, organized, and analyzed when conducting the internal and external assessments and when making informed predictions about the future, organizations make decisions about the most appropriate strategic emphasis and the best competitive strategy to adopt.

With these decisions made, organizations use the information now available to them to develop a written strategic plan that contains the following elements:

1. Vision statement
2. Mission statement
3. Guiding principles
4. Strategic goals

Once the strategic plan is completed, it must be executed. A strategic plan that is developed and then just placed on a shelf to gather dust is nothing more than a written summary of some good ideas. To help the organization gain competitive advantage in the marketplace, a strategic plan must be executed, and the more effective the execution the more competitive the organization.

This book sets forth a ten-step strategic planning model developed by the author and used effectively for more than 30 years with organizations ranging in size from large corporations to small "mom-and-pop" shops. The model presented in this book has been revised, refined, and enhanced over more than 30 years based on experience and application. The model as presented herein represents a comprehensive approach to strategic planning that is convenient and easy to use, but thorough and effective. The ten steps, each of which is explained in detail in the chapters that follow, are:

1. Understand the concept of competition and competitive advantage.
2. Understand the concept of *value* as it relates to competitive advantage.
3. Know the rationale for strategic planning.
4. Understand the strategic planning process.
5. Collect, organize, and analyze applicable data.
6. Make informed predictions about the future.
7. Establish direction and formulate strategy.
8. Write the strategic plan.
9. Execute the plan.
10. Understand how the process fits together from start to finish.

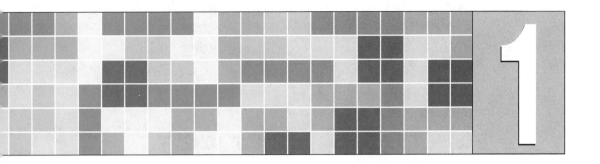

Understand the Concepts of Competition and Competitive Advantage

Competitive advantage is at the heart of a firm's performance in competitive markets. Firms throughout the world face slower growth as well as domestic and global competitors that are no longer acting as if the expanding pie were big enough for all.

—Michael E. Porter, *Competitive Advantage*

OBJECTIVES

- Define the concept of competition as it relates to technology companies.
- Define the concept of competitive advantage as it relates to technology companies.
- Explain the ramifications of global competition.

- Explain the universal forces that determine the potential profitability of an industry.
- Explain the characteristics of globally competitive companies.
- Explain the advantages and disadvantages of U.S. companies in the global marketplace.

COMPETITION AND THE TECHNOLOGY COMPANY

Competition is the process by which organizations attempt to establish and sustain a profitable position by outperforming other organizations in the same markets. Said another way, competition is what happens when two or more organizations are trying to achieve the same result, and only one of them can win. For example, when two football teams line up across from each other on the field, they both want to win the game, but only one of them can. Hence they compete.

The ability to compete in a global marketplace has become synonymous with the ability to survive in business. Remember, the global marketplace includes both domestic and foreign competitors. Every functional unit in a technology company, every strategy the company adopts, every individual the company employs should contribute in some way to making the company more competitive. If a functional unit does not make an organization more competitive, why have it? If a strategy does not make an organization more competitive, why adopt it? If an individual does not make an organization more competitive, why employ him or her?

In a global environment, a company might seek several possible competitive positions (Figure 1.1). The overall goal of any competitive position is sustained profitability. The ideal competitive position is to have a large share of a large market created by human needs that are so fundamental the market will sustain itself for the long term (Figure 1.1a). Sustainability is the key here. A large share of a large but temporary market will not lead to sustained profitability no matter how substantial the one-time profits may be. Large markets tend to be those that grow out of the most fundamental human needs or that appeal to the most common human interests. If the market is large enough and long term in nature, even a relatively small share can be sufficient for sustained profitability (Figure 1.1b).

One should not assume that small markets—those that grow out of less fundamental needs or interests—cannot yield sustained profitability. A large enough share of a relatively small market can be sufficient for success at a certain level (Figure 1.1c). The competitive position a technology company does not want is a small share of a small market (Figure 1.1d).

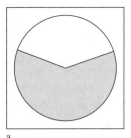

a.
Ideal:
Large share of a large market

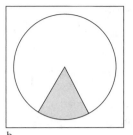

b.
Acceptable:
Small share of a large market

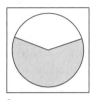

c.
Acceptable:
Large share of a small market

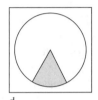

d.
Questionable Value:
Small share of a small market

FIGURE 1.1

Fundamentally Necessary Industries

Many factors combine to determine an organization's ability to compete. If sustained profitability over the long term is the desired result of a given competitive position, one of the most critical determinants will be the market potential of the industry in question. In general terms, the industries with the most potential are those that respond to the most fundamental human needs. Several factors addressed later in this chapter can affect the profitability of a given industry, even in those markets that on the surface would appear to have the most potential. For now, let's focus on potential.

On the surface, the potential for sustained profitability appears strong in industries that provide goods and services so closely tied to the daily lives and work of people—that is to say, they are so **fundamentally necessary**—that

---

Strategic Planning Tip

Key Point

If sustained profitability over the long term is the desired result of a given competitive position, one of the most critical determinants will be the market potential of the industry in question.

Strategic Planning Tip

Key Point

Since the market in fundamentally necessary industries is so large, even securing a relatively small share of the market might be sufficient to provide an acceptable level of profitability. Said another way, even a small piece of a large enough pie can be quite filling.

they will always be needed or will be needed as far into the future as it is possible to determine. Such industries as transportation (e.g., automobiles, trucks, aircraft, watercraft), durable goods (e.g., refrigerators, stoves, freezers, washing machines, dryers), consumer electronics (e.g., computers, televisions, DVD players, stereos, cameras), and military hardware (e.g., aircraft, ships, tanks, vehicles, artillery, munitions) are examples of industries so fundamentally necessary that they are likely to have long-term potential in their respective markets. Producers of electricity, natural gas, oil and gasoline, and water are also in fundamentally necessary industries.

Fundamentally necessary industries have the potential for sustained profitability, but entering such an industry is no guarantee of success. Companies in even the most fundamentally necessary industries can still fail if they cannot perform effectively enough to gain and maintain a sufficient market share. In addition, determinants of profitability covered later in this chapter can limit the *real* potential of an industry that on the surface has *apparent* potential. On the other hand, since the market in fundamentally necessary industries is so large, even securing a relatively small share of the market might be sufficient to provide an acceptable level of profitability. Said another way, even a small piece of a large enough pie can be quite filling. To illustrate this point, consider how many automobile makers there are in the world and their respective shares of the overall market. Many automobile makers exist on the basis of a relatively small share of a huge global market.

Less Fundamentally Necessary Industries

Some industries provide products and services for markets that are less fundamentally necessary. Their products and services are more **discretionary** in nature. Without considering the determinants covered later in this chapter, these industries appear to have less potential for sustained profitability over the long term than do the more fundamentally necessary industries.

Industries that produce products or provide services tied into the recreation, leisure, and vacation travel markets fall into this category. During

Strategic Planning Tip

Key Point

To compete successfully in a global environment, technology companies must establish *competitive advantage.*

peak economic times, discretionary products and services do well. However, during recessions these markets can fall off precipitously. For example, many of the industries tied into the recreation, leisure, and vacation travel markets struggled to survive following the terrorist attacks of September 11, 2001.

The ideal competitive position for a company is to be a high performer in a high potential industry. However, just as being in an industry with high potential for sustained profitability is no guarantee of success, being in an industry with less potential is no guarantee of failure. Companies in lower-potential industries can actually do quite well if they can outperform their competitors. Regardless of whether the industry in question has high or low apparent potential for sustained profitability, the level of success enjoyed will be tied directly to the concept of competitiveness. To compete successfully in a global environment, technology companies must establish **competitive advantage.**

COMPETITIVE ADVANTAGE AND THE TECHNOLOGY COMPANY

Whether the market in question is large or small, fundamental or discretionary, broad or niche, a prerequisite to gaining competitive advantage in that market is to provide *superior value.* Some customers focus primarily on the cost of the product or service provided. Some place more value on certain attributes of the product or service and, as a result, are willing to pay more for those attributes. In reality, value to a given customer is always some combination of cost and desired attributes. Depending on circumstances, one customer might place more value on low cost while another might place more value on selected attributes. To gain competitive advantage, a technology company must understand its customers well enough to know what they want and how they define value.

Gaining competitive advantage is a matter of providing the best value at the lowest cost—the **low-cost strategy**—or of providing such superior value in terms of attributes that customers are willing to pay more for the product or service. The latter approach is called the **differentiation strategy.** A technology company differentiates its products and services by

<hr>

Strategic Planning Tip

Key Point

Gaining competitive advantage is a matter of providing the best value at the lowest cost—the *low-cost strategy*—or of providing such superior value in terms of attributes that customers are willing to pay more for the product or service. The latter approach is called the *differentiation strategy*.

<hr>

building attributes into them that make the products and services more valuable to customers than are those of competitors.

Low-end automobiles, sometimes called *economy cars*, are products of a low-cost strategy. This does not mean that product attributes such as quality and reliability are not important to people who purchase economy cars. They are. Rather, it means that because of their economic circumstances or for other reasons, the highest priority of people who purchase economy cars is cost. Such people buy the best car they can get for the price they can afford. The makers of economy cars apply the low-cost strategy and hope that a higher volume of sales will result in sustained profitability.

High-end cars are products of a differentiation strategy. This does not mean that cost is not a factor for people who purchase more expensive cars. Rather, it means they place so high a priority on such attributes as quality, reliability, comfort, and image that they are willing to pay the higher cost to get them. It also means they can afford to value attributes over cost.

Competitive Strategies[1]

The two basic types of competitive strategy—low cost and differentiation— can each be either focused in scope or broad in scope. As a result, technology companies can adopt one of four competitive strategies:

1. Low-cost/broad scope
2. Differentiation/broad scope
3. Low-cost/narrow scope (niche)
4. Differentiation/narrow scope (niche)

- *Low-cost/broad scope.* With this competitive strategy, a technology company attempts to be the low-cost provider of certain products or services to a broad and varied customer base. This strategy requires companies to produce highly standardized, bare-bones products that have a generic appeal while maintaining the most efficient processes and the lowest possible cost relationships with suppliers.

Strategic Planning Tip

Key Point

The two basic types of competitive strategy—low cost and differentiation—can each be either focused in scope or broad in scope.

- *Differentiation/broad scope.* With this competitive strategy, a technology company attempts to differentiate itself by providing products and services with attributes so valuable to customers that they are willing to pay a higher price for them. This is how Federal Express originally differentiated itself from the U.S. Postal Service. Guaranteed overnight delivery, the ability to track packages, and in-office pickup were attributes many customers were willing to pay extra for and that the Postal Service did not provide at that time. The keys to success with this strategy are (1) identifying attributes that are different from those of competitors and highly valued by customers and (2) pricing the product or service high enough to cover the costs associated with providing them.
- *Low-cost/narrow scope.* This is the low-cost strategy applied to a limited segment of a broader market. It could also be called the *low-cost niche strategy.* To succeed with this strategy, a company must identify a segment of a broader customer base with buying behavior that differs from that of the broader base. By focusing on that segment and tailoring its design, production, delivery, and services, a company can be the low-cost niche provider and still achieve sustained profitability. For example, in the field of computer-aided design and drafting or CAD systems, the broad market includes architectural, civil engineering, structural engineering, mechanical engineering, aeronautical engineering, electrical engineering, manufacturing, and construction applications. A producer of CAD systems that chooses to hold down the price of its product by focusing

---

Strategic Planning Tip

Key Point

Although World War II caused people to think globally, advances in technology really made the global marketplace possible. Advances in electronic communication such as the Internet made people from all over the world electronic neighbors and electronic customers.

narrowly and defining its market niche as just civil engineering applications is applying the low-cost/narrow scope competitive strategy.

- *Differentiation/narrow scope.* This competitive strategy is a variation of the differentiation strategy. The principal difference is that with the narrow scope, the technology company focuses on a more limited set of attributes and differentiates them even more than competitors that differentiate for a broader market. As a result, both the customer base and the list of product attributes are narrower than those of companies pursuing a differentiation strategy that is broader in scope.

RAMIFICATIONS OF GLOBAL COMPETITION

World War II and subsequent technological advances in the fields of transportation and electronic communication created the global marketplace. Following the war, industrialized countries began looking for markets outside their own borders. In an odd way, the war had brought the world closer together. Although World War II caused people to think globally, advances in technology really made the global marketplace possible. Advances in electronic communication such as the Internet made people from all over the world electronic neighbors and electronic customers. Advances in transportation technology made the rapid and reliable delivery of products to virtually any location on the globe not just a possibility, but a daily occurrence.

The Berlin Airlift following World War II proved that large quantities of goods could be quickly moved by air to any location in the world. This meant that raw materials produced in one country could be used in the manufacture of products in a second country that, in turn, are used by consumers in third, fourth, and fifth countries. For example, leather produced in Australia might be shipped as raw material to Italy where it is used in the manufacture of shoes and purses sold to consumers in the United States, France, and Japan. At the same time, leather produced in South America might be sent to shoe manufacturers in Indonesia. These manufacturers, like their Italian counterparts, sell their shoes to consumers in the United States, France, and Japan. This means that the manufacturers in Italy compete with the manufacturers in Indonesia. This simplified example demonstrates the kind of competition that takes place on a global scale every day. Such competition has become the norm, and it can be intense.

At one time only large multinational corporations faced global competition. Now even small companies are affected. Today no company is immune to the effects of global competition. A company doing business in a small town in Middle America might find itself competing with companies in Indonesia, Malaysia, or China.

STRATEGIC PLANNING PROFILE	Competitive Advantage at Southwest Airlines and United Parcel Service

A critical part of the strategic planning process involves deciding how the organization will gain a competitive advantage in the marketplace. The organization's overall competitive strategy should be made clear in the written strategic plan. The strategic plans of Southwest Airlines Company and United Parcel Service (UPS) show how these two well-known companies attempt to gain and maintain competitive advantage:

Competitive Strategy of Southwest Airlines Company
Southwest Airlines Company applies the following strategies to gain competitive advantage: low-cost fares, high customer satisfaction, and focused marketing (business commuters and leisure travelers). This is a low-cost, narrow scope strategy that also attempts to differentiate on the basis of customer satisfaction.

Competitive Strategy of United Parcel Service
UPS applies the following strategies to gain competitive advantages in the global marketplace: competitive rates, innovative solutions to global delivery needs, efficient distribution worldwide, excellent service, and rapid response to changing conditions. This is a differentiation/broad scope strategy that also attempts to gain the benefit of competitive rates.

Source: Jeffrey Abrahams, *The Mission Statement Book* (Ten Speed Press, Berkeley, CA: 1999).

UNIVERSAL DETERMINANTS OF INDUSTRY PROFITABILITY[2]

Earlier in this chapter the concept of *apparent potential* for sustained profitability was explained. This section shows how to take the next step and determine the *actual potential* for sustained profitability. The factors explained in this section can show that an industry with even the highest apparent potential has in reality little or no potential for sustained profitability.

The point has been made that sustained profitability is the goal of a company's competitive strategy. If the playing field is level, the company with the best competitive strategy that most effectively carries out that strategy will win in the global business arena. But, of course, the playing field in global business is seldom level. Just how level the playing field is can determine the potential profitability of a given industry. Michael E. Porter sets forth the following five forces that shape the playing field in both the domestic and global business arenas:

- Entry of new competitors
- Threat of substitutes

Key Point

If the playing field is level, the company with the best competitive strategy that most effectively carries out that strategy will win in the global business arena. But, of course, the playing field in global business is seldom level.

- Bargaining power of buyers
- Bargaining power of suppliers
- Rivalry among existing competitors

Together, these forces determine the potential profitability of a given industry by determining if and to what extent the playing field is level. Before moving into a given industry, a technology company needs to know the ramifications of these five forces as they relate to that industry. The following example illustrates how these factors can affect a company's profitability in a given industry. Assume that a technology company is considering becoming a new entrant in an existing market. Before deciding whether or not to take the plunge, the company wants to conduct a cost/benefit analysis. That analysis will focus on Porter's five forces. How much **bargaining power** do suppliers have? This will affect how much it costs to bring the new product to market. The more bargaining power the suppliers have, the more they can charge for their raw materials and inputs. How much bargaining power do customers have? If several competitors are already producing the same or a similar product (intensity of rivalry), buyers might have sufficient bargaining power to hold down the selling price of new entrants.

If the company cannot charge enough for its product to generate an acceptable return on its investment, why enter the new market in the first place? If there is a high propensity among buyers to substitute products, this must also be weighed as an issue during the cost/benefit analysis. The types of questions a company might ask when applying Porter's five forces are outlined in the following paragraphs.

Key Point

If a given industry is attractive—that is, it appears to have high profitability potential—there will always be the threat of new competitors entering the field.

Entry of New Competitors

If a given industry is attractive—that is, it appears to have high profitability potential—there will always be the threat of new competitors entering the field. Every time this happens, both buyers and suppliers can, at least potentially, gain bargaining power. In addition, the intensity of the rivalry among existing competitors is likely to increase. How then can a technology company decide if it should be a new entrant into a given market? Porter lists the following factors to consider:

1. *Economies of scale.* Have existing competitors established economies of scale that will make it difficult or impossible to compete with them? Can our company establish economies of scale that will help us outperform existing competitors?

2. *Proprietary product differences.* Do existing competitors have proprietary rights to product attributes or conversion processes that allow them to differentiate their products in ways our company will not be able to overcome? Does our company have proprietary rights that will allow us to differentiate our products sufficiently to gain an acceptable market share?

3. *Brand identity.* Have existing competitors established brand identity to the point of exclusion? Can our company use its existing brand identity from other products to break into this new market?

4. *Capital requirements.* Are the capital requirements to break into the new market acceptable or prohibitive? Are we comfortable that our company can quickly gain sufficient market share to overcome the capital requirements of bringing the new product to market and still show an acceptable return on investment within an acceptable timetable?

5. *Access to distribution.* Do competitors have established distribution networks that will make it difficult or impossible to compete? Does our company have access to distribution networks that will make us competitive in getting our new product out to the market?

6. *Absolute cost advantages.* Do existing competitors have any irrevocable cost advantages that will always give them a competitive advantage? Does our company have any permanent cost advantages that will help us break into the new market and gain an acceptable market share?

7. *Government policy.* Do existing government policies favor entrenched competitors, or are there new or pending policy changes that might open the door for new competitors?

8. *Expected retaliation.* How can we expect existing competitors to react if we enter this market? Will they improve their product in ways that will make it even more difficult to gain a foothold in this market?

Will they lock in orders with suppliers that lock our company out? Will they retaliate by introducing products to compete with our existing products in other markets? Is our company prepared and able to counter any negative retaliation? Is the market share our company might gain by becoming a new entrant worth the cost of fending off retaliatory moves by existing competitors?

Threat of Substitutes

Before entering a new market, a technology company should consider the potential its new product will have as a substitute for existing products in the market in question. Will the market accept a substitute for existing products? What is the relative pricing of our substitute product versus the existing competitors' products? What are the costs associated with switching to our substitute? Do the price advantages of our company's substitute product offset the cost customers will incur in switching?

Bargaining Power of Buyers

A technology must be able to gain a sufficient market share within a reasonable time frame in order to show an acceptable return on the investment (ROI) it made breaking into a new market. One factor that will have a major impact on ROI potential is the bargaining power of customers. On the one hand, the more bargaining power customers have, the more pressure there is to hold down prices. On the other hand, the ability to apply an appropriate markup on its products is one of the critical keys in generating acceptable ROI within a reasonable time frame. High bargaining power on the part of customers can be a major barrier to entering a new market.

Porter lists the following determinants of buyer bargaining power: buyer concentration compared with competitor concentration (the more providers of the same or similar products, the more power the buyer has), buyer volume, buyer switching costs, buyer information, substitute products, product differences, brand identity, impact on quality and performance, buyer profits, and incentives for decision makers.

Bargaining Power of Suppliers

Whereas the bargaining power of buyers is a major determinant of the price a technology company can charge for its product, the bargaining power of suppliers is a major determinant of what the company must pay for its raw inputs. The more power suppliers have, the more they can charge for their inputs. This in turn drives up the price the company must charge for its product. Porter lists the following determinants of supplier power: differentiation of input, availability of substitute inputs, supplier concentration, importance of volume to the supplier, and the impact of supplier inputs on cost or differentiation strategies.

Rivalry Among Existing Competitors

How much rivalry already exists among competitors in the market? A technology company considering becoming a new entrant in an existing market must answer this question. If the rivalry is already intense—if competitors are already investing the resources needed to keep their products differentiated—it will be difficult to break into the market. Will a new entrant cause existing rivals to increase the level of competition? Porter lists the following determinants of the intensity of competitive rivalry: industry growth, fixed costs/value added, intermittent overcapacity, product differences, brand identity, switching costs, concentration and balance, informational complexity, diversity of competitors, corporate stakes, and exit barriers.

CHARACTERISTICS OF GLOBALLY COMPETITIVE COMPANIES

It is often said that only "world-class" companies can compete in the global marketplace. But what is a world-class organization? In an attempt to answer this question, the American Management Association (AMA) conducted a global survey of more than 2,000 managers in 36 different countries. The AMA received 1,797 usable responses from 36 countries. According to this survey, the following are the top 15 areas in which organizations are concerned about doing well as they attempt to compete in the global marketplace:

1. Customer service
2. Quality
3. Research and development/new product development
4. Acquiring new technologies
5. Innovation
6. Team-based approach (adopting and using effectively)
7. Best practices (study and use of)
8. Manpower planning
9. Environmentally sound practices
10. Business partnerships and alliances
11. Reengineering of processes
12. Mergers and acquisitions
13. Outsourcing and contracting
14. Reliance on consulting services
15. Political lobbying

In addition to these issues, the AMA survey also found that respondents were concerned about several human resource topics. The ten most important of these are:

1. Worker productivity (improvement)
2. Employee training and development
3. Open communication between management and employees
4. Employee benefits and perquisites
5. Codes of workplace conduct
6. Conflict resolution
7. Employee satisfaction
8. Flextime arrangements
9. Management-employee-union relations
10. Child care[3]

U.S. COMPANIES IN THE GLOBAL MARKETPLACE[4]

As business continues the current trend toward globalization, how are companies in the United States faring? A business trying to compete in the global marketplace is like an athlete trying to compete in the Olympics. Nowhere is the competition tougher. Correspondingly, no country in the world gives its businesses such a solid foundation from which to work. According to Weidenbaum, the following six factors account for a country's ability to compete in the international marketplace:[5]

1. An economy that is open to foreign investment and trade
2. A government that minimizes controls on business, and does a good job of supervising financial institutions
3. A judicial system that works well and helps reduce corruption
4. Greater transparency and availability of economic information
5. High labor mobility, which enhances productivity and thus living standards
6. Ease of entry by new businesses

In varying degrees, the United States meets all six of these criteria. Of course, how well these criteria are fulfilled is a matter of debate between and among various interest groups and stakeholders. Nonetheless, when compared with other countries competing in the global marketplace, the United States fares well in all six of these key areas. This being the case, a key advantage of American firms trying to compete in the global marketplace is these six factors working in their favor. Other advantages and disadvantages are summarized in the following sections.

Global Advantages of U.S. Companies

In the global marketplace, the United States is the world leader in the following industries: aerospace, airlines, beverages, chemicals, computer services, electrical products, entertainment, general merchandise, motor vehicles, office equipment, paper products, pharmaceuticals, photographic and scientific equipment, semiconductors, soap and cosmetics, and tobacco. Some of the reasons the United States leads the world in these key industries are the following advantages:

1. Strong entrepreneurial spirit
2. Presence of a "small capitalization" stock market for small and midsized companies
3. Rapidly advancing technologies
4. Comparatively low taxes
5. Low rate of unionization
6. World-class system of higher education (colleges and universities)

The United States leads the world in new business start-ups. This is because the entrepreneurial spirit is an integral part of the American persona. The presence of a small capitalization stock market allows small and midsized companies to start up and expand without having to use all of their own capital or to take out higher interest loans from banks, as is often the case in other countries. The United States leads the world in the development, transfer, diffusion, and use of technology. This helps ensure a continual stream of new products on the one hand and improved productivity on the other. Americans complain constantly about taxes (as they are entitled to do in practicing their rights as free citizens), but when compared with other industrialized nations, the United States has a low tax burden. Tension between labor and management can harm productivity and, in turn, decrease a company's ability to compete in the global marketplace. The amount of tension that exists between labor and management can typically be demonstrated by the level of union activity: the more tension, the more union activity. Compared with other industrialized nations, union activity in the United States is low.

The United States also provides the world's best higher education system. The number of top-notch colleges and universities in the United States is so much greater than those in other countries that comparisons are irrelevant. The cost of higher education in America, although viewed as high by U.S. citizens, is inexpensive when compared with that of other industrialized nations. In addition, financial aid is so readily available that almost any person with the necessary academic ability can pursue a college education in the United States. In addition, America's community college system—the gateway to higher education for so many—is a uniquely American concept.

Global Disadvantages of U.S. Companies

In spite of the many strengths companies in the United States can bring to the global marketplace, and in spite of this country's world-leading position in several key industries, companies must deal with some disadvantages. The primary global disadvantages of U.S. companies are:

1. Expanding government regulations
2. A growing "underclass" of have-nots
3. A weak public school system (K-12)
4. A poorly skilled labor force and poor training opportunities
5. An increasing protectionist sentiment (to restrict imports)
6. Growing public alienation with large institutions (public and private)

Regardless of which major political party has controlled Congress over the past 25 years, the general trend has been toward increasing government regulation of business. Regulating business is a difficult balancing act. On the one hand, businesses cannot be allowed to simply pursue profits, disregarding the potential consequences to consumer safety, the environment, and other national interests. On the other hand, too much regulation or unnecessary regulation can make it impossible to compete globally. The growing divide between "haves" and "have-nots" in the United States might lead to the establishment and perpetuation of a permanent economic and social underclass. This is precisely what happened in Russia when Czar Nicholas II was overthrown by the Communists in the early 1900s. People who lose hope might very well respond in ways that threaten the nation's peace, stability, and social fabric. One of the key factors in the establishment of a social and economic underclass is the failure of America's public school system (K-12). Even with the best system of higher education in the world, America cannot overcome the shortcomings of its K-12 system. In fact, if drastic improvements are not made, over time those shortcomings will begin to erode the quality of our higher education system.

The most fundamental problem with the public school system from the perspective of global competition is that most of the jobs in companies that need to compete globally require less than a college education. These jobs must be performed by high school graduates who, if they cannot read, write, speak, listen, think, and calculate better than their counterparts in other countries, will be outperformed. Poorly skilled workers are an outgrowth of the failure of the nation's public school system, in which the overwhelming majority of Americans are educated. Ideally, every high school graduate should be fully prepared to either go to work or go on to college. When this is not the case, as it certainly is not, American companies must try to compete with a less skilled labor force. This is like a base-ball coach trying to win with a team of players who cannot pitch, catch, run, or hit.

One of the factors that contributed to the Great Depression of the 1930s was global protectionism. Americans wanted their farmers and their manu-facturers to be "protected" from their counterparts in other countries. Pro-tectionism hurts everyone and never really protects anyone. But as other countries have entered U.S. markets (principally Japan, Korea, and China), the jobs of American workers have been threatened. A natural but ill-informed response is to call for protectionist measures and to adopt slogans such as "Buy American." Economists are quick to point out, however, that the only valid reason to "buy American" is that American products have the most value. If they do not, buying them makes little sense and is nothing more than misguided patriotism. The better approach is to ask why the American products are not the best and then to do what is necessary to make them the best.

The final factor that puts U.S. companies at a disadvantage is the grow-ing tendency of the public to see big organizations as the "bad guys." This is displayed in many different ways. Disgruntled employees sometimes pre-tend injuries and file fraudulent workers' compensation claims. Employees cheat and steal from their employers. Of course the most common way em-ployees act out animosity toward big business is by giving less than their best on the job. Another expression is when the public at large supports an-tibusiness legislation and unnecessary regulations.

Summary

1. Competition is the process by which organizations attempt to estab-lish and maintain a profitable position by performing better than other or-ganizations in the same market. Sustained profitability is the goal of the competitive strategies organizations adopt.

2. Generally speaking, those industries with the greatest potential for sus-tained profitability are those that are the most fundamentally necessary in terms of basic human needs. There is at least the potential for sustained prof-itability in industries that provide goods and services closely tied to the daily lives and work of people. On the other hand, the fact that a given industry has high potential for sustained profitability is no guarantee of success. Com-panies in the highest potential industries can still fail if they cannot perform effectively enough to gain and maintain sufficient market share.

3. Whether the market in question is large or small, fundamental or dis-cretionary, broad or niche, a prerequisite to gaining competitive advantage in that market is to provide superior value.

4. Gaining competitive advantage is a matter of providing the greatest value at the lowest cost—the *low-cost strategy*—or of providing such superior value in terms of attributes that customers are willing to pay more for the product or service. The latter approach is called the *differentiation strategy*.

5. Organizations can adopt one of four generic competitive strategies: low-cost/broad scope, differentiation/broad scope, low-cost/narrow scope, and differentiation/narrow scope.

6. The ramifications of global competition are that no company is immune to the hyper-competitive stresses of international competition. Even the smallest firms that deal only in domestic markets now find themselves being challenged by competitors from other countries.

7. Michael E. Porter sets forth the following determinants of an industry's profitability potential: entry of new competitors, threat of substitutes, bargaining power of buyers, bargaining power of suppliers, and rivalry among existing competitors.

8. The American Management Association (AMA) identified the following characteristics of globally competitive companies: customer service, quality, research and development/new product development, acquiring new technologies, innovation, team-based approach, study and use of best practices, manpower planning, environmentally sound practices, business partnerships and alliances, reengineering of processes, mergers and acquisitions, outsourcing and contracting, reliance on consulting services, and political lobbying.

9. Weidenbaum identified the following advantages of U.S. companies in the global marketplace: strong entrepreneurial spirit, small capitalization for small and mid-sized companies, rapidly advancing technologies, comparatively low taxes, low rate of unionization, and world-class system of higher education.

10. Weidenbaum identified the following disadvantages of U.S. companies in the global marketplace: expanding government regulation, growing underclass of have-nots, weak public school system, poorly skilled labor force, increasing protectionist sentiments, and growing alienation with large institutions.

Key Terms and Concepts

Absolute cost advantages	Brand identity
Access to distribution	Capital requirements
Bargaining power of buyers	Competition
Bargaining power of suppliers	Competitive advantage

Differentiation strategy

Differentiation/broad scope

Differentiation/narrow scope

Discretionary industries

Economies of scale

Entry of new competitors

Expected retaliation

Fundamentally necessary industries

Global advantages of U.S. companies

Global disadvantages of U.S. companies

Government policies

Low-cost strategy

Low-cost/broad scope

Low-cost/narrow scope

Proprietary product differences

Rivalry among existing competitors

Threat of substitutes

Review Questions

1. Define the concept of competition as it relates to technology companies.
2. What are fundamentally necessary industries?
3. What are discretionary industries?
4. Explain the concept of competitive advantage.
5. How does the concept of superior value affect an organization's ability to compete?
6. Compare and contract the low-cost/broad scope and the low-cost/narrow scope competitive strategies.
7. Compare and contrast the differentiation/broad scope and differentiation/narrow scope competitive strategies.
8. List and explain the universal determinants of industry profitability set forth by Porter.
9. List the characteristics of globally competitive companies.
10. According to Weidenbaum, what are the advantages of U.S. companies competing in the global marketplace? What are the disadvantages?

SIMULATION CASES FOR DISCUSSION

The following simulation cases are provided to generate additional thought and discussion about the principles explained in this chapter. Readers are encouraged to consider how the situations presented in these cases might apply to them and to discuss the cases with others interested in strategic planning and execution.

CASE 1.1   Should We Pursue This New Market or Not?

A debate was taking place in the CEO's office of Trans-Tech, Inc., (TTI) concerning a proposal to enter a new market. The vice president for marketing wanted TTI to develop a substitute product and attempt to break into what he sees as a "huge" market. The vice president for engineering disagreed, claiming there are already too many entrenched competitors in the market in question.

"So, what you are saying is that you think even a small share of this market would be worth the effort," said the CEO to his marketing vice president.

"Exactly," responded the marketing executive. "This pie is so big that even a small piece of it will be big enough."

"I still disagree," snorted the engineering executive. "If we can't gain at least a 20 percent market share, what is the point of even going into this business?"

Discussion Questions

1. Join this debate. Who do you agree with in this debate and why?
2. Can you think of a company that has succeeded by maintaining a relatively small share of a large market?

CASE 1.2   What Competitive Strategy Should We Adopt?

The executives from TTI in Case 1.1 decided to move forward with their new product. Their next debate concerned the issue of competitive strategy.

"I think we should pursue a low-cost strategy for a specific market niche," offered the director of manufacturing. "I know we can produce our substitute at less cost than the major competitors in this market."

"We can now," agreed the vice president for engineering. "But what happens if we lose a couple of your key players? Without Jones and Welch our productivity and quality will decrease by 20 percent."

"He's right," said the CEO. "I favor the niche market concept, but would rather adopt a differentiation strategy."

Discussion Questions

1. What do you see as the ramifications of each proposal in this case—low-cost niche or differentiation niche?
2. Can you name a company that maintains an acceptable market share using the low-cost niche strategy? How about the differentiation niche strategy?

`CASE 1.3` Aren't We Forgetting Something Here?

The executives of TTI from Case 1.2 decided to adopt a differentiation niche strategy and were just about ready to close the discussion and get to work when the company's chief financial officer said, "Aren't we forgetting something here?"

"What do you mean?" asked the CEO.

"I mean that before we go rushing off to develop this new product and adopt a competitive strategy, shouldn't we first do a cost/benefit analysis? We haven't even considered the bargaining power of buyers and suppliers or the potential threat of retaliation from existing competitors."

"He's got a point," said the CEO.

Discussion Questions

1. Do you agree that there is still another step to complete in the process before making a final decision? Explain your rationale.

2. How might the various determinants of industry profitability affect TTI's decision? Give examples.

Endnotes

1. Porter, M. E., *Competitive Advantage* (The Free Press, New York: 1985), 11–16.

2. Ibid. 4.

3. American Management Association, "AMA Global Survey on Key Business Issues," *Management Review* (December 1998), 27–38.

4. DeFeo, J. A., "The Tip of the Iceberg," *Quality Progress*, 34 (5), (May 2001), 29–37.

5. Weidenbaum, M., "All the World's a Stage," *Management Review* (October 1999), 44.

Understand the Concept of Value As It Relates to Competitive Advantage

We never know the value of water until the well is dry.[1]

—Thomas Fuller

OBJECTIVES

- Define the concept of *value* as it relates to competitive advantage.
- Explain the interrelated components of the value tree.
- Describe how organizations can know what their customers value.
- Explain the rationale for continually improving value.
- List the essential value improvement activities.
- Explain how to structure an organization for continual improvement.

- Provide an overview of the scientific approach to continual improvement of value.

- Describe the basic strategies for improving value-producing processes.

As we found in Chapter 1, an organization looking for an industry or a new market to enter should first consider the apparent potential for sustained profitability of that industry or market. Then, having found an industry or market that appears, at least on the surface, to be attractive, the organization should use the five determinants of profitability set forth by Porter[2] to conduct a detailed analysis to uncover the real potential of the industry or market in question. If, after due consideration, an industry or market is determined to have real potential for sustained profitability, the company may choose to move forward.

At this point the organization's performance as compared with that of its competitors becomes a critical issue. To win in the global business arena, the organization must provide greater *value* to customers than do other organizations targeting the same customers. Whether the competitive strategy chosen is low cost or differentiation, a key ingredient in success will be providing greater value than the competition.

Strategic Planning Tip

Key Point

Value is simply the price people are willing to pay for the product or service they want.

This chapter is about value and how organizations can provide superior value to customers. The material presented here will help readers understand the concept of value as it relates to establishing and maintaining a competitive advantage in the marketplace as well as how an organization's functions and processes work together to produce superior value.

VALUE DEFINED

Value is an easy concept to define but can be a complex concept to grasp. **Value** is simply the price people are willing to pay for the product or service they want. The complexity in understanding this concept lies in the fact that different people place a different value on the same product and service characteristics. In other words, like beauty, value is in the eyes of the beholder. Consequently, one of the keys to understanding value is in knowing your customers well enough to know what they value.

Some customers place more value on cost while others place more value on product or service attributes. But all customers attach a certain amount of value to both cost and attributes. Value is seldom an either–or proposition. It is a combination of both cost and desired attributes. Which factor turns out to be more important to a given customer depends to a great extent on a variety of factors, including customer preferences, economic status or condition, and personality. In fact, value is as much a psychological concept as it is a business concept. To complicate matters even further, what a given customer values at a given time can change as his or her circumstances change.

This complication can be illustrated using a simple example. When John was working his way through college as an engineering student, he was always broke. Just paying tuition, room, board, and book costs was an ongoing challenge. Consequently, when John's old clunker of a car finally broke down and he faced the daunting prospect of having to find room in his budget for a car payment, his highest priority was cost. Later, when John was established as a practicing engineer and doing well financially, he decided it was time to buy a new car. This time, because his circumstances had changed, John placed a higher priority on such attributes as reliability, safety rating, architecture (e.g., body style, color, interior features), and engine size. Because he now could afford to do so, John placed a higher priority on product attributes than cost. Cost was still a concern, but not nearly the concern it was when John was a struggling college student with perpetually empty pockets.

Customers are just like John regardless of whether they are individual consumers, other companies, or government agencies. A struggling start-up company is likely to place more value on cost than will a well-established company that is better off financially. However, as the start-up company's financial condition improves, it might begin to focus more attention on product attributes.

THE VALUE TREE

Assume that a technology company has selected an industry with satisfactory potential for sustained profitability. Assume also that the company has adopted a strategy that has the potential to make it competitive in the chosen industry. The key from this point on will be to provide superior value to customers. To illustrate how this is done, let's consider the analogy of a tree.

Figure 2.1 is what the author calls the **value tree.** Companies that compete in the global business arena are like trees. To develop strong branches and healthy foliage, a tree must be firmly rooted, and it must be able to efficiently and effectively process the raw inputs (air, sunlight, and nutrients) brought into it. To provide products and services of superior value, technology companies must be firmly rooted by their support functions (e.g., human resources, acquisitions, and new product development), and they must be able to efficiently and effectively convert raw inputs into quality products and services.

The human resources function provides the people with the knowledge, skills, and experience the organization needs to do business. The acquisition function purchases the raw materials that will be converted into products and services. The new product development function responds to the

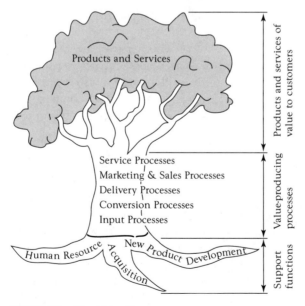

FIGURE 2.1   The Value Tree

Strategic Planning Tip

Key Point

Just as a tree that efficiently and effectively processes good air, water, and nutrients will produce quality foliage, a technology company that has efficient and effective value-producing processes will provide products and services of value for its customers.

---

ever-changing market to ensure that the company keeps up with what customers want.

The trunk of the value tree consists of five categories of processes known collectively as the **value-producing processes.** Ensuring peak performance of these processes is essential to competitiveness regardless of the competitive strategy adopted. The more efficiently and effectively the value-producing processes operate, the more competitive the company will be. Outperforming competitors who rely on the same or similar processes is one of the best ways to gain a competitive advantage in the global business arena. If the company in question were a baseball team, its value-producing processes would be pitching, hitting, fielding, and base stealing. If the company were a football team, the value-producing processes would be blocking, tackling, running, and passing. To be globally competitive, a technology company must establish performance base lines for all of its value-producing processes, benchmark the processes against best-in-class performers, and continually improve the performance of these processes.

For a technology company, the value-producing processes include the following types of processes in each broad category:

1. *Input processes.*   Handling and inspecting incoming materials and parts; sorting, inventorying, and storing incoming materials and parts; and delivering materials and parts to the appropriate work centers on either a traditional or just-in-time basis.

2. *Conversion processes.*   Fabrication processes used to convert raw materials and parts into finished products and services (e.g., forming, casting, machining, joining, assembly, quality control, maintenance, repair, retrofitting, and plant operations).

3. *Delivery processes.*   Order taking, packaging/containerizing, and shipping processes.

4. *Marketing and sales processes.*   Promotion, advertising, customer visitation, communication, and sales processes.

5. *Service processes.*   Product installation, maintenance, repair, retrofitting, and spare parts processes.

STRATEGIC PLANNING PROFILE   Providing Superior Value at Ferro Corporation

Ferro Corporation is a global company that provides ceramic glaze and porcelain enamel coating products for manufacturing applications. The company also produces electronic materials, tile systems, and PVC specialty additives. To thrive in a global environment, Ferro Corporation must provide superior value to its customers. The company views the following as being competitive advantages that allow it to provide superior value:

- "Our extensive global presence allows us to serve the needs of multinational customers efficiently and leverage our technology and product development skills."
- "Our strong R&D capabilities mean innovative product and process solutions for our customers. Core technologies link our businesses and product lines and provide synergies."
- "Our vast technical expertise helps customers meet exacting specifications and improve their own product and operations. We not only know how to capitalize on industry trends, we have a hand in shaping many of them."

Source: Ferro Corporation Web site (http://www.ferro.com).

Just as a tree that efficiently and effectively processes good air, water, and nutrients will produce quality foliage, a technology company that has efficient and effective value-producing processes will provide products and services of value for its customers. The company that continually improves the performance of its value-producing processes will provide *superior* value for its customers.

KNOW WHAT CUSTOMERS VALUE: COLLECTING CUSTOMER INPUT

Every company has several potentially productive sources of customer information, but the information may not be readily apparent. Like ore, it may need to be mined from these sources: warranty records, customer service records, frontline personnel, and customer interviews (see Figure 2.2).

Warranty Records

Companies that guarantee their products, services, and work have a built-in source of customer information. When a product is returned for repair or replacement, a record is made. When a crew must be dispatched to correct warranted work that did not hold up for the required period of time or

**Sources of
Customer Information**

- Warranty Records
- Customer-Service Records
- Frontline Personnel
- Customer Interviews

FIGURE 2.2   Look in-house for customer information.

to repair equipment that does not function properly, a record is made. These records, if thoroughly analyzed, contain a wealth of information on customer needs and attribute preferences.

Attributes are features of a product or characteristics of a service that customers value. For example, most automobile buyers think reliability is an important attribute. They want to know that every time they put the key in the ignition and turn it, the car will start promptly. A car that is hard to start is not reliable and, if under warranty, will surely cause complaints and produce repair work. In such cases, records are kept and can be examined to identify frequent problems or problem trends.

As another example, customers who deal with engineering firms expect to receive accurate, well-prepared drawings. Accuracy is an attribute expected of engineering plans, calculations, and bills of material. Inaccuracies in any of these products lead to problems that, in turn, lead to unplanned work for which there will be records.

Using warranty records to identify product attributes and service characteristics that fail to meet customer expectations would seem to be an obvious strategy. However, many companies simply file warranty records away and never look at them again. Companies that fail to regularly analyze their warranty records ignore a potential treasure chest of customer information.

Customer Service Records

Closely related to warranty records are customer service records. Companies that provide customers a hotline to call or an e-mail address to contact when they have problems have an excellent source of customer information on which to draw. Customer service calls, unlike warranty complaints, are not necessarily about product failures or workmanship discrepancies. Often they go like this: "I just purchased your product, and I cannot get it to work. What am I doing wrong?" Software manufacturers frequently receive this

---

Key Point

Companies that fail to regularly analyze their warranty records ignore a potential treasure chest of customer information.

---

kind of call. In such cases, the product will probably operate as designed, but only if customers know how to do their part. The problem could be insufficient or poor customer training, a poorly written user manual, or poorly designed operating procedures. Even the best-designed product is faulty if customers cannot figure out how to use it.

Frontline Personnel

Frontline personnel are those employees who have the most contact with customers. They typically include receptionists, clerical personnel, secretaries, sales personnel, marketing representatives, and after-purchase service providers. Personnel in these categories have access to a wealth of customer information, but too few companies regard them as legitimate sources of customer information. The term *mining customer information* is used to describe the process of collecting such information from frontline personnel. This is because, like mining, the process involves digging out the information. It can take a concerted effort. Collecting customer information from frontline personnel is no passive undertaking. It means engaging frontline personnel in periodic data collection efforts (e.g., internal focus groups, one-on-one conferences, internal surveys). The process should be regular and systematic.

Customer Interviews

You can learn a lot about customers' needs and preferences by just sitting down and talking with them. There are two approaches to conducting customer interviews: the walk-in approach and the by-invitation approach. The walk-in approach is used when a customer takes the initiative to make an appointment or just drops in. In either case, customers have a problem they want to discuss. The meeting is useful for more than just hearing the complaint. Once you have used the appropriate assertive listening strategies, you transform the meeting into an interview. This is done simply by asking customers if they would mind answering some questions to help you improve customer satisfaction.

Strategic Planning Tip

Key Point

The term *mining customer information* is used to describe the process of collecting such information from frontline personnel. This is because, like mining, the process involves digging out the information.

The by-invitation approach differs primarily in that the company takes the initiative to ask customers for an interview. The customers are given advance notice of the types of questions the company representative plans to ask. This gives them an opportunity to prepare, thereby making better use of everyone's time. Another important difference between the walk-in format and the by-invitation format is the number of customers interviewed. With the walk-in format, customers are taken one by one, as they appear. In essence, the walk-in customer is told, "You were coming anyway. Since you are here, why not help us out by answering some questions?" With the by-invitation format, the number of interviews arranged depends on how many are necessary to adequately cover all segments of the company's market(s). For example, an engineering firm that provides services in more than one engineering field should interview several customers in each field. If a manufacturer produces more than one product, customers from all market segments should be interviewed. A comprehensive sample will yield the most accurate results.

Regardless of whether the interview is initiated by invitation or on a walk-in basis, its purpose is to collect information to be analyzed further and acted upon appropriately by company officials.

Strategic Planning Tip

Key Point

You can learn a lot about customers' needs and preferences by just sitting down and talking with them.

RATIONALE FOR CONTINUALLY IMPROVING VALUE

Continual improvement is fundamental to success in the global marketplace. A company that is just maintaining the status quo in such key areas as new product development, the adoption of new technologies, and process performance is like a runner who is standing still in a race. Competing in the global marketplace is like competing in the Olympics. Last year's records are sure to be broken this year. Athletes who don't improve continually are not likely to remain long in the winner's circle. The same is true of companies that must compete globally.

Customer needs are not static; they change continually. A special product feature that is considered innovative today will be routine tomorrow. A product cost that is considered a bargain today will be priced too high to compete tomorrow. A good case in point in this regard is the ever-falling price for each new feature introduced in the personal computer. The only way a company can hope to compete in the global marketplace is to improve continually.

ESSENTIAL VALUE IMPROVEMENT ACTIVITIES

Continual improvement is not about solving isolated problems as they occur. Such an approach is just "putting out fires." Solving a problem without correcting the fault that caused it—in other words, simply putting out the fire—means the problem will occur again. Peter R. Scholtes sets forth the following activities as being crucial to continual improvement:[3]

- *Maintain communication.* Communication is essential to continual improvement. This cannot be overemphasized. It is important to share information before, during, and after attempting to make improvements. All people involved as well as any person or unit that might be affected by a planned improvement should know what is being done, why, and how it might affect them.

- *Correct obvious problems.* Often process problems are not obvious, and a great deal of study is required to isolate them and find solutions. This is the typical case and it is why the scientific approach is so important in making process improvements. However, sometimes a process or product problem will be obvious. In such cases, the problem should be corrected immediately. Spending days studying a problem for which the solution is obvious just so that the scientific approach is used will result in $100 solutions to $10 problems.

Strategic Planning Tip

Key Point

Competing in the global marketplace is like competing in the Olympics. Last year's records are sure to be broken this year. Athletes who don't improve continually are not likely to remain long in the winner's circle. The same is true of companies that must compete globally.

- *Look upstream.* Look for causes, not symptoms. This is a difficult point to make with people who are used to taking a cursory glance at a situation and "putting out the fire" as quickly as possible without taking the time to determine what caused it.

- *Document problems and progress.* Take the time to write it down. It is not uncommon for an organization to continue solving the same problem over and over again because nobody took the time to document which problems have been dealt with and how they were solved. A fundamental rule of continual improvement is "document, document, document."

- *Monitor changes.* Regardless of how well studied a problem is, the solution eventually adopted may not solve it or may only partially solve it, or it may produce unintended consequences. For this reason, it is important to monitor the performance of a process after changes have been implemented. It is also important to ensure that pride of ownership on the part of those who recommended the change does not interfere with objective monitoring of the changes. These activities are essential regardless of how the improvement effort is structured.

Strategic Planning Tip

Key Point

Continual improvement is not about solving isolated problems as they occur. Such an approach is just "putting out fires."

STRUCTURE FOR CONTINUAL IMPROVEMENT

Continual improvement doesn't just happen. It must be undertaken in a systematic, step-by-step manner. For an organization to make continual improvements, it must be structured appropriately. Structuring for continual improvement involves the following steps:[4]

- *Establish a quality council.* The quality council has overall responsibility for continual improvement. According to Juran, "The basic responsibility of this council is to launch, coordinate, and 'institutionalize' annual quality improvement."[5] It is essential that the membership include executive-level decision makers.

- *Develop a statement of responsibilities.* All members of the quality council, as well as employees who are not currently members, must understand the council's responsibilities. One of the council's first priorities is to develop and distribute a statement of responsibilities bearing the signature of the organization's CEO. Responsibilities that should be stated include the following: (a) formulating policy as it relates to continual improvement; (b) setting the benchmarks; (c) establishing the team and project selection processes; (d) providing the necessary resources, such as training and time away from job duties to serve on a project team; (e) implementing projects; (f) establishing measures for monitoring progress and undertaking monitoring efforts; and (g) implementing an appropriate reward and recognition program.

■ *Establishing the necessary infrastructure.* The quality council constitutes the foundation of an organization's continual improvement effort. However, there is more to the infrastructure than just the council. The remainder of the infrastructure consists of subcommittees of the council assigned responsibility for the specific duties, project improvement teams, training, and a structured improvement process.

THE SCIENTIFIC APPROACH TO CONTINUAL IMPROVEMENT

The scientific approach is fundamental to continual improvement. It involves using facts to make decisions and identifying root causes rather than treating symptoms. Using the scientific approach involves applying the following four steps:

1. Collect factual information.
2. Identify root causes.
3. Develop solutions based on facts.
4. Implement solutions and adjust.

Collect Factual Information

It is not uncommon for an organization or a unit within it to collect unverified data or to make a procedural error that results in the collection of erroneous data. In fact, in the age of computers, this is quite common. Decisions based on meaningless or erroneous data are bound to lead to failure. Before collecting data, decide exactly what is needed, how it can best be collected, where the data exists, how it will be measured, and how you will know the data are accurate. Then convert the data into factual in-

---

Strategic Planning Tip

Key Point

Decisions based on meaningless or erroneous data are bound to lead to failure.

formation by formatting it in ways that apply directly to situations in question.

Identify Root Causes

Identifying root causes is critical. Too many resources are wasted by organizations attempting to solve symptoms rather than problems. Unless the root cause of the problem is eliminated, the problem will recur.

Develop Solutions Based on Facts

Never assume a given solution. Collect the relevant data, make sure it is accurate, identify root causes, and then develop a solution that is appropriate. Acting on inaccurate assumptions is a sure way to waste resources.

Plan and Implement Solutions

Planning forces you to look ahead, anticipate needs and what resources will be available to satisfy them, and anticipate problems and consider how they should be handled. Much of the scientific approach has to do with establishing reliable performance indicators and using them to measure actual performance. Giorgio Merli lists the following as examples of useful performance indicators for value-producing processes:[6]

- Number of errors or defects
- Number of or level of need for repetitions of work tasks
- Efficiency indicators (units per hour, items per person)
- Number of delays
- Duration of a given procedure or activity
- Response time or cycle
- Usability/cost ratio
- Amount of overtime required
- Changes in workload
- Vulnerability of the system
- Level of criticalness
- Level of standardization
- Number of unfinished documents

This is not a complete list. Many other indicators could be added. Those actually used vary widely from organization to organization. However, such indicators, regardless of which ones are actually used, are an important aspect of the scientific approach.

Key Point

Much of the scientific approach has to do with establishing reliable perform-ance indicators and using them to measure actual performance.

BASIC STRATEGIES FOR IMPROVING VALUE-PRODUCING PROCESSES

Numerous value-producing processes are used in business and industry; consequently, there is no single road map to follow when improving processes. However, a number of standard strategies can be used as a menu from which improvement strategies can be selected as appropriate. Basic generic strategies are explained in the following sections.[7]

Describe the Process

The strategy of describing a process is used to make sure that everyone in-volved in improving it has a detailed knowledge of the process. Usually this requires some investigation and study. The steps involved are as follows:

1. Establish boundaries for the process.
2. Flowchart the process—as it is, not as it should be.
3. Make a diagram of how the work flows.
4. Verify your work.
5. Correct immediately any obvious problems identified.

Standardize the Process

To continually improve a process, all people involved in its operation must be using the same procedures. Often this is not the case. Employee X may use different procedures than Employee Y. It is important to ensure that all employees are using the best, most effective, most efficient procedures known. The steps involved in standardizing a process are as follows:

1. Identify the currently known best practices and write them down.
2. Test the best practices to determine whether they are, in fact, the best, and improve them if there is room for improvement (these improved practices then become the final best practices that are recorded).

3. Make sure that everyone is using the newly standardized process.

4. Keep records of process performance, update them continually, and use them to identify ways to improve the process even further on a continual basis.

Eliminate Errors in the Process

The strategy of eliminating errors in the process involves identifying errors that are commonly made in the operation of the process and then getting rid of them. This strategy can help delete steps, procedures, and practices that are being done a certain way simply because that is the way they have always been addressed. Whatever measures can be taken to eliminate such errors are carried out as part of this strategy.

Streamline the Process

The strategy of streamlining the process is used to take the slack out of a process. This can be done by reducing inventory, reducing cycle times, and eliminating unnecessary steps. After a process has been streamlined, every step in it has significance, contributes to the desired end, and adds value.

Reduce Sources of Variation

The first step in the strategy of reducing sources of variation is identifying the sources. Such sources can often be traced to differences among people, machines, measurement instruments, material, sources of material, operating conditions, and times of day. Differences among people can be attributed to levels of capability, training, education, experience, and motivation. Differences among machines can be attributed to age, design, and maintenance. Regardless of the source of variation, after a source has been identified, this information should be used to reduce the amount of variation to the absolute minimum. For example, if the source of variation is a difference in the levels of training completed by various operators, those who need more training should receive it. If one set of measurement instruments is not as finely calibrated as another, they should be equally calibrated.

Bring the Process under Statistical Control

The strategy of bringing the process under statistical control requires in-depth knowledge of the concept of statistical process control (SPC). A control chart is planned, data are collected and charted, special causes are eliminated, and a plan for continual improvement is developed.

Improve the Design of the Process

There are many different ways to design and lay out a process. Most designs can be improved. The best way to improve the design of a process is through an active program of experimentation. To produce the best results, an experiment must be properly designed using the following steps:

1. Define the objectives of the experiment. (What factors do you want to improve? What specifically do you want to learn from the experiment?)
2. Decide which factors to measure (e.g., cycle time, yield, finish).
3. Design an experiment that will measure the critical factors and answer the relevant questions.
4. Set up the experiment.
5. Conduct the experiment.
6. Analyze the results.
7. Act on the results.

Summary

1. Value is the price people are willing to pay for the product or service they want. Like beauty, value is in the eyes of the beholder. Consequently, one of the keys to understanding value is to know your customers well enough to know what they value.

2. The value tree is an analogy that describes how technology companies can compete in the global business arena. To have strong branches and healthy foliage, a tree must be firmly rooted, and it must efficiently and effectively process its raw inputs (water, air, sunlight, and nutrients) brought into it. To provide products and services of superior value, technology companies must be firmly rooted by their support functions (e.g., human resources, acquisitions, and new product development). Their value-producing processes must efficiently and effectively process this raw input into finished products and services.

3. The value-producing processes for technology companies fall into the following categories: input, conversion, delivery, marketing and sales, and service processes.

4. Technology companies can learn much about what customers want and what they value by examining warranty records, mining customer service records, talking with frontline personnel, and conducting customer interviews.

5. Continual improvement is fundamental to success in the global marketplace. A company that is just maintaining the status quo is like a runner

standing still in a race. Customer needs are not static; they change constantly. Consequently, the only way a technology company can hope to compete in the global marketplace is to improve continually.

6. Essential improvement activities include: maintaining communication, correcting obvious problems, looking upstream, documenting problems and progress, and monitoring changes.

7. Technology companies should structure themselves for continual improvement by doing the following: establishing a quality council, developing a statement of responsibilities for the quality council, and establishing the necessary infrastructure for continual improvement.

8. When making improvements, it is important to use the scientific approach. This involves identifying root causes of problems and collecting factual information about those causes. The steps are as follows: collect factual information, identify root causes, develop solutions based on facts, and implement the solutions and adjust as necessary.

9. Basic strategies for improving value-producing processes include the following: describe the process, standardize the process, eliminate errors in the process, streamline the process, reduce sources of variation, bring the process under statistical control, and improve the design of the process.

Key Terms and Concepts

Bring the process under statistical control

Collect factual information

Conversion processes

Correct obvious problems

Customer interviews

Customer service records

Delivery processes

Describe the process

Develop solutions based on facts

Document problems and progress

Eliminate errors in the process

Frontline personnel

Identify root causes

Implement solutions and adjust

Improve the design of the process

Input processes

Look upstream

Maintain communication

Marketing and sales processes

Monitor changes

Reduce sources of variation

Service processes

Standardize the process

Streamline the process

Value tree

Value

Value-producing processes

Warranty records

Review Questions

1. Define the concept of value as it relates to competitive advantage.
2. Explain all of the various components of the value tree.
3. How can an organization know what its customers value?
4. What is the rationale for continually improving value?
5. List and describe the essential value improvement activities.
6. How can a technology company structure itself for continual improvement?
7. Describe how to apply the scientific approach to the concept of continual improvement.
8. Explain the basic strategies for improving value-producing processes.

SIMULATION CASES FOR DISCUSSION

The following simulation cases are provided to generate additional thought and discussion about the principles explained in this chapter. Readers are encouraged to consider how the situations presented in these cases might apply to them and to discuss the cases with others interested in strategic planning and execution.

CASE 2.1 Do We Know Our Customers Well Enough?

Penny Cotter, CEO of MapTech, Inc., was talking with her senior leadership team about how they can hold onto their company's market leadership.

"Can anyone tell me why our customers purchase our services instead of those of our competitors?" She went around the conference table and asked each executive to answer her question. There were a variety of answers, and some were good, but all were obviously just speculation.

"Those are all good guesses," said Cotter. "But that's all they really are—guesses based on anecdotal evidence. I don't think we really know our customers well enough. Next time we meet, I'd like to hear your recommendations concerning how we can learn what our customers really want—what they value the most about our products and services."

Discussion Questions

1. What are some ways companies pick up anecdotal evidence concerning what their customers value about their products and services?

2. Put yourself in this scenario. At the next meeting, what recommendations would you make to the CEO?

CASE 2.2 We Are Already the Best—Why Try to Improve on Success?

Once Penny Cotter, CEO of MapTech, Inc., had collected the recommendations from her senior executives concerning how they could get to know their customers' wants, needs, and preferences better, she implemented "Operation Customer Input." Cotter received a summary report of the findings of Operation Customer Input, but her senior executives had not seen the report.

"Today I want to discuss ways in which we can improve our products and services," said Cotter.

"You aren't serious, are you?" asked the vice president for engineering. "We are already the best in our market. Why try to improve on success?"

Discussion Questions

1. Do you agree or disagree with the vice president for engineering? His argument can be summed up as *don't fix what isn't broken.*
2. If you were Penny Cotter, how would you respond to the vice president for engineering's comment?

CASE 2.3 I Agree We Need to Improve Continually— How Do We Get Started?

After reading the summary report Penny Cotter gave them at the previous meeting, MapTech's senior executives now understood that even though their company's products and services were currently the best on the market, they still needed to be improved and improved continually. The discussion at the next meeting was about what to do next.

"I'd like to hear your opinions on how to get started making improvements," said Cotter. "I want our improvement efforts to be scientific, systematic, and ongoing. I want continual improvement."

Discussion Questions

1. If you were part of MapTech's senior executive team, how would you explain the application of the scientific approach to continual improvement?
2. What basic strategies would you recommend the company implement?

Endnotes

1. Tripp, R. T. (compiled by), *The International Thesaurus of Quotations* (Thomas Y. Crawell Co., New York: 1970), 1019.8.

2. Porter, M. E., *Competitive Advantage* (The Free Press, New York: 1988), 5.

3. Scholtes, P. R., *The Team Handbook* (Joiner Associates, Madison, WI: 1992), 5–6 through 5–9.

4. Juran, J., *Juran on Leadership for Quality: An Executive Handbook* (The Free Press, New York: 1989), 72.

5. Ibid., 43.

6. Merli, G., *Total Manufacturing Management* (Productivity Press, Cambridge, MA: 1990), 143.

7. Scholtes, 5–54 through 5–67.

Know the Rationale for Strategic Planning

The top priority of management is to ensure that there's a shared vision for the future of the organization and that this vision is part of the culture.

—Robert W. Bradford and J. Peter Duncan, *Simplified Strategic Planning*

OBJECTIVES

- Explain why it is important to develop a comprehensive, written strategic plan.

- Explain how the strategic planning process can promote buy-in to the organization's strategic direction.

- Describe the appropriate membership of an organization's strategic planning team.

- Explain critical strategies for leading the strategic planning team.

Many organizations develop strategic plans, but in the author's experience, too many fail to use them. Here is a typical scenario. A strategic plan is developed, printed with impressive graphics, and bound in an attractive cover. The organization's vision, mission, and guiding principles (sometimes called *core values*) are engraved on an expensive plaque and hung on the wall in the organization's lobby. All of this is good as far as it goes. Unfortunately, all too often this is as far as it goes. With many organizations, once the framed copy is hung on the lobby wall, the strategic plan is put on a shelf to gather dust and be forgotten.

A strategic plan is, or at least should be, the most important management tool in an organization's toolbox. Unfortunately, with many organizations the strategic plan is a neglected tool. Rather than being used to guide the organization to victory in the global business arena, the strategic plan sits rusting at the bottom of the toolbox. An organization that develops a strategic plan and then fails to use it is like a contractor who has a set of plans for the house he is building, but never bothers to look at them.

Strategic Planning Tip

Key Point

Establishing direction for an organization causes stakeholders to look into the future, and this is a good thing to do. However, strategic planning is not just about looking into the future—it's about shaping the future.

IMPORTANCE OF DEVELOPING A WRITTEN STRATEGIC PLAN

Would you want to embark on a vacation cruise with a ship's captain who failed to chart a course, direction, and schedule? Probably not. Would you want to embark on a business trip with a pilot who failed to file a flight plan? Again, probably not. People involved in such endeavors like to know where they are going, what it will take to get there, and when they will arrive. More importantly, they like to know that those who will pilot the ship or airplane know where they are going and when they will get there. What is true of people embarking on vacations and business trips is true of people in organizations. They want to know where their organization is going and how they will know when it has arrived. This is the **direction rationale** for strategic planning (see Figure 3.1).

Establishing direction for an organization causes stakeholders to look into the future, and this is a good thing to do. However, strategic planning is not just about looking into the future—it's about shaping the future. A National Football League team goes into the Super Bowl with a game plan

---

**Rationale for
STRATEGIC PLANNING**

- Direction
- Future-shaping
- Big picture
- Beacon in the distance
- Stakeholder buy-in

FIGURE 3.1   Reasons for developing a strategic plan.

not just for playing the game, but for winning it. To win the Super Bowl, a team must do more than just show up and leave the game's outcome to fate. It must take responsibility for its own destiny and shape the outcome of the game. This is the purpose of its game plan—to shape the game's outcome. The same may be said of an organization's strategic plan. Its purpose is to help the organization take charge of its own destiny and shape its own future. This is the **future-shaping rationale** for strategic planning.

Another rationale for strategic planning is to create a context within which an organization operates. This is called the **big-picture rationale** for strategic planning. It is important for all personnel in an organization to understand the big picture concerning what their organization is all about and where it is going. It is also important for them to understand where they fit into that big picture. People in an organization can become so compartmentalized by their individual responsibilities within their respective functional units that they lose sight of their organization's overall purpose and direction (big picture), or never understand it in the first place. In addition, the everyday pressure of the job—schedules, deadlines, budgets, turnover, personnel problems, etc.—can distract stakeholders from what really matters.

Strategic Planning Tip

Key Point

A comprehensive, written strategic plan that is reviewed regularly will serve the purpose of reminding personnel of the big picture and where they fit into it. Personnel who understand the big picture are better equipped to play a positive role in moving the organization in the right direction.

Strategic Planning Tip

Key Point

When people are involved in the planning process, the product of that process—the strategic plan—becomes their plan rather than your plan.

A comprehensive, written strategic plan that is reviewed regularly will serve the purpose of reminding personnel of the big picture and where they fit into it. Personnel who understand the big picture are better equipped to play a positive role in moving the organization in the right direction. Those who understand the big picture will also understand how best to use their limited resources to serve the organization's best interests when choices must be made about the application of resources.

Another rationale for developing a comprehensive, written strategic plan can be found in the **beacon analogy**. Just as a lighthouse provides a beacon in the distance to guide ships through uncharted waters or foggy conditions, a strategic plan provides a beacon in the distance to keep stakeholders on course when they enter on a new course of action or when decisions become engulfed in a fog of confusion. It can be difficult to make good decisions when the mind is clouded by the exigencies of the daily grind. In such cases, the strategic plan serves as a beacon that shines through the fog and helps stakeholders stay on course toward the organization's ultimate destination.

Stakeholder buy-in is another rationale for developing a comprehensive, written strategic plan. When personnel buy into a plan, they will support it as their own. It becomes "their plan" rather than a plan that was inflicted on them. In this case, the strategic planning process, if properly handled, and not the plan itself generates buy-in. How to ensure that the planning process has this desired effect is covered in the next section.

PROMOTING BUY-IN THROUGH THE STRATEGIC PLANNING PROCESS

Stakeholders are more likely to help carry out a plan if they buy into it. Buy-in means more than just accepting the strategic plan, it means believing in it and wanting to make it a reality. Consequently, it is critical that organizations achieve buy-in from all stakeholders when developing a strategic plan. The planning process as much as the plan itself offers the most potential for achieving buy-in.

When people are involved in the planning process, the product of that process—the strategic plan—becomes *their* plan rather than *your* plan. The charitable organization Habitat for Humanity applies this principle when

Strategic Planning Tip

Key Point

Among the most reassuring of human endeavors are those in which people proactively attempt to take control of their own destinies. The strategic planning process has the effect of giving people a feeling of taking control of where they and their organization are going.

it builds a new house for a low-income family who cannot afford to buy a home. Habitat personnel know from experience how important it is for the new owners to buy into the house and think of it as their own rather than as the charitable gift of some nice volunteers. Habitat wants its clients to feel like *homeowners*, not recipients of charity. They achieve this goal by involving their clients in actually helping build their new home. By investing their own "sweat equity" in building their new home, Habitat's clients buy into the project, and it becomes a partnership between them and the Habitat volunteers rather than a charity project. An organization's strategic plan is like a Habitat for Humanity house. If you want stakeholders to own it, let them help build it.

The importance of the concept of buy-in is found in the benefits that accrue when an organization is successful in securing it. These benefits are as follows:

1. *Sense of purpose.*   The strategic planning process gives those involved a sense of purpose. Consequently, it is important to ensure that the planning process is broad-based. In other words, the broader the involvement of personnel, the broader the buy-in to the plan.

2. *Enthusiasm.*   Just as children who are asked by their parents to help plan the family vacation show more enthusiasm for the vacation, personnel who help develop an organization's strategic plan will show more enthusiasm for the plan.

3. *Commitment.*   People involved in shaping their future are more likely to commit to that future than those who feel the future is something that is forced upon them.

4. *Control.*   People don't like to feel that events are controlling them. Among the most reassuring of human endeavors are those in which people proactively attempt to take control of their own destinies. The strategic planning process has the effect of giving people a feeling of taking control of where they and their organization are going.

5. *Stress.*   Feeling out of control and subject to the random exigencies of fate is stressful for people. The strategic planning process can relieve

---

Strategic Planning Tip

Key Point

The composition and leadership of the strategic planning team will be critical determinants of the quality of the completed plan.

---

the stress associated with feeling out of control and helpless in terms of the future.

6. *Focus.*   The strategic planning process gives stakeholders focus. Through their participation in the process, stakeholders come to understand the big picture and where they fit into it. This allows them to focus their attention, efforts, and resources and know that their focus is appropriate.

7. *Motivation.*   The strategic planning process can be an effective tool for motivating personnel to give their best efforts every day and to continually improve what is considered their best effort. Once people understand where their organization is trying to go, and once they buy into the planned direction, they are more likely to be self-motivated to do their parts to help the organization reach the planned destination.

THE STRATEGIC PLANNING TEAM

To produce a good strategic plan, an organization must put together a good strategic planning team. The composition and leadership of the strategic planning team will be critical determinants of the quality of the completed plan. These factors will also help determine the extent to which the organization will enjoy the buy-in benefits explained in the previous section.

Strategic Planning Tip

Key Point

One or two known creative thinkers should be added to the membership of the strategic planning team to ensure a steady flow of new ideas, challenge the status quo, and offer alternatives to conventional thinking.

Three questions must be answered before establishing an organization's strategic planning team:

1. Who should be on the team?
2. How many team members should there be?
3. Who should lead the team?

Team Membership

The strategic planning team should consist of individuals who meet one or more of the following criteria: (1) have the authority to commit the organization to a given direction; (2) have the ability to think creatively; (3) have the self-confidence to interact with executives without being intimidated; and (4) have the ability to represent a specific constituency.

The organization's senior executives are the only individuals who meet the first criterion of authority. Therefore, the CEO and the individuals who make up his or her senior management team should be the nucleus of the strategic planning team. To meet the second criterion—creative thinking— an organization has a great deal of latitude. Most organizations have on staff individuals who are known for thinking more creatively than their colleagues. One or two known creative thinkers should be added to the membership of the strategic planning team to ensure a steady flow of new ideas, challenge the status quo, and offer alternatives to conventional thinking.

It is important to note here that creative thinkers often ruffle the feathers of people who think conventionally. Their ideas are not always welcomed and their opinions are not always valued. On a strategic planning team, those getting their feathers ruffled might be the organization's senior managers. Consequently, any creative thinker put on the team must meet the third criterion—self-confidence. No person who is cowed in the presence of senior executives or is easily intimidated should be a member of the planning team.

Concerning the fourth criterion—specific constituencies—the CEO and senior executives should attempt to ensure that all applicable perspectives in the organization are represented on the planning team. This criterion is satisfied in large measure by making the senior management team the

Strategic Planning Tip

Key Point

A good rule of thumb is to establish a team with enough members to ensure that all constituencies in the organization are represented, but not so many members that the team is too cumbersome to manage.

nucleus of the strategic planning team. Senior managers, by definition, will encompass all of the organization's functional divisions, departments, and units and, as a result, will ensure that their perspectives are represented. The one remaining perspective that should be considered is that of employees (labor). Whether the organization is unionized or not, it is always a good idea to have a member of the team who can represent the perspective of frontline employees.

Size of the Strategic Planning Team

The size of the team will vary according to the size and complexity of the organization. A good rule of thumb is to establish a team with enough members to ensure that all constituencies in the organization are represented, but not so many members that the team is too cumbersome to manage. Most organizations can pull together an effective team with anywhere from six to ten members. Don't be concerned about having an even number of team members. Strategic planning teams are not democratic bodies. They don't vote. Rather, they work for consensus.

Are fewer than six members acceptable? Yes, as long as the team can adequately cover all functional aspects of the organization, represent a variety of perspectives, and ensure an appropriate mix of creativity and conventionality. Are more than ten members acceptable? Yes, as long as the team is manageable. Obviously, there are no magic numbers that are automatically right when deciding on the composition of the strategic planning team. Putting together a team that will be effective requires a great deal of discussion, careful thought, and good judgment.

Strategic Planning Tip

Key Point

An effective way to broaden the base of involvement in strategic planning is to task each member of the planning team with establishing one or more ad hoc committees to complete specific duties relating to plan development.

Broadening Involvement Beyond the Planning Team

The strategic planning team will consist of six to ten carefully selected individuals. Even in a small company, this does not represent broad-based involvement. The various members of the team do represent all constituent groups in the organization, but being involved and being represented are not the same thing. Organizations that do the best jobs of developing and using strategic plans involve many stakeholders beyond just the members of the strategic planning team.

An effective way to broaden the base of involvement in strategic planning is to task each member of the planning team with establishing one or more ad hoc committees to complete specific duties relating to plan development. Ad hoc committees can be tasked with conducting research as part of the organization's analysis of its strengths and weaknesses, for example. They can be used as sounding boards for bouncing ideas off of, as feedback mechanisms for identifying discrepancies in the thinking of the planning team, and as resources for mining new ideas and creative solutions.

Strategic Planning Tip

Key Point

Leadership is the single most important ingredient in achieving effective work in teams. With good leadership, ordinary people can achieve extraordinary results.

TEAM LEADERSHIP

Leadership is the single most important ingredient in achieving effective work in teams. With good leadership, ordinary people can achieve extraordinary results. On the other hand, even the most talented people will produce mediocre results if they are poorly led. To produce a good strategic plan, the team must be well led.

Necessary Abilities of Team Leaders

In their book *Leaders: The Strategies for Taking Charge*, Warren Bennis and Burt Nanus describe three lessons for leadership that summarize what leaders must be able to do.[1]

1. *Overcome resistance to change.* Some people attempt to do this using power and control. Leaders overcome resistance by achieving a total, willing, and voluntary commitment to shared values and goals.

2. *Broker the needs of constituency groups inside and outside of the team.* When the needs of the team and another unit appear to conflict, leaders must be able to find ways of bringing the needs of both together without shortchanging either one.

> **Strategic Planning Tip**
>
> ## Key Point
>
> Good team leaders are good communicators. They are willing, patient, skilled listeners. They can also communicate their ideas clearly, succinctly, and in a nonthreatening manner. They use their communication skills to establish and nurture rapport with team members.

3. *Establish an ethical framework within which all team members and the team as a whole operate.* This is best accomplished by doing the following:

   - Setting an example of ethical behavior
   - Choosing ethical people as team members
   - Communicating a sense of purpose for the team
   - Reinforcing appropriate behaviors within the team and outside of it
   - Articulating ethical positions, internally and externally

What Is a Good Team Leader?

Good team leaders are able to inspire people to make a total, willing, and voluntary commitment because they have the characteristics shown in Figure 3.2.

Good leaders are committed to both the job to be done and the people who must do it, and they can strike an appropriate balance between the two. Good leaders project a positive example at all times. They are good role models. Managers who project a "Do as I say, not as I do" attitude are not effective team leaders. To inspire employees, team leaders must be willing to do what they expect of workers—and do it better, do it right, and do it consistently. If, for example, dependability is important, team leaders must set a consistent example of dependability. If punctuality is important, team leaders must set a consistent example of punctuality. To be a good leader, one must set a consistent example of all characteristics that are important on the job.

Good team leaders are good communicators. They are willing, patient, skilled listeners. They can also communicate their ideas clearly, succinctly, and in a nonthreatening manner. They use their communication skills to establish and nurture rapport with team members. Good leaders have influence with employees and use it in a positive manner. *Influence* is the art of using power to move people toward a certain end or point of view. The power of team leaders derives from the authority that goes with their jobs

STRATEGIC PLANNING PROFILE   **Anheuser Busch Companies**

The three critical elements in a company's strategic plan are the vision, mission, and guiding principles or values. An excellent example of these elements can be found in the strategic plan of the Anheuser Busch Companies:

Vision

"Through all of our products, services and relationships, we will add to life's enjoyment."

Mission

"Be the world's beer company."

"Enrich and entertain a global audience."

"Deliver superior returns to our shareholders."

Values

"Quality in everything we do."

"Exceeding customer expectations."

"Trust, respect and integrity in all of our relationships."

"Continuous improvement, innovation and embracing change."

"Teamwork and open, honest communication."

"Each employee's responsibility for contributing to the company's success."

"Creating a safe, productive and rewarding work environment."

"Building a high-performing, diverse workforce."

"Promoting the responsible consumption of our products."

"Preserving and protecting the environment and supporting communities where we do business."

Source: Material in quotes comes from the Anheuser Busch Web site (http://www.anheuser-busch.com)

and the credibility they establish. Power is useless unless it is converted to influence. Power that is properly, appropriately, and effectively applied becomes positive influence.

Finally, good leaders are persuasive. Team leaders who expect people simply to do what they are told have limited success. Those who use their communication skills and influence to persuade people to their point of view and to convince them to make a total, willing, and voluntary commitment to that point of view can have unlimited success.

Checklist of
Traits of Good Team Leaders

___✓___ Balanced commitment between the work and team members

___✓___ Positive role model for team members

___✓___ Good communication skills

___✓___ Positive influence on team members

___✓___ Persuasiveness with team members

FIGURE 3.2   Good team leaders have these traits.

Team Leadership Styles

Leadership styles are based in how leaders interact with those they lead. Leadership styles go by many different names, but most fall into one of the following categories: autocratic, democratic, participative, goal-oriented, and situational.

- *Autocratic leadership.*   Autocratic leadership is also called *directive* or *dictatorial leadership.* People who take this approach make decisions without consulting the employees who will have to implement them or who will be affected by them. They tell others what to do and expect them to comply. Critics of this style say that although it can work in the short run or in isolated instances, it is not effective in the long run. Autocratic leadership is not appropriate in a team setting.

- *Democratic leadership.*   Democratic leadership is also called *consultative leadership.* People who take this approach involve stakeholders who will have to implement the decision. After receiving the input and recommendations of team members, the leader takes a vote. Critics of this approach say the most popular decision is not always the best decision and that democratic leadership, by its nature, can result in popular decisions as opposed to right decisions. This style can also lead to compromises that ultimately fail to produce the desired result. Democratic leadership is not appropriate in a team setting.

- *Participative leadership.*   Participative leadership is also known as *nondirective leadership.* People who take this approach exert reduced control over the decision-making process. They provide information about the problem and ask team members to develop strategies and solutions. The underlying assumption of this style is that workers more readily accept responsibility for solutions, goals, and strategies if they help develop them. Critics of this approach say it is time consuming and works only if all people involved are committed to the best interests of the team.

■ *Goal-oriented leadership.* Goal-oriented leadership is also called *results-based leadership.* People who take this approach ask team members to focus solely on the goals at hand. Only those strategies that make a definitive and measurable contribution to accomplishing team goals are discussed. The influence of personalities and other factors unrelated to specific team goals are minimized. Critics of this approach say it can break down when team members focus so intently on specific goals that they overlook opportunities or potential problems that fall outside their narrow focus. These drawbacks can make results-based leadership ineffective in team settings.

■ *Situational leadership.* Situational leadership is also called *contingency leadership.* People who take this approach select the leadership style that seems appropriate based on the circumstances that exist at a given time. In evaluating these circumstances, leaders consider the following factors:

  ■ The relationship of the leader and team members
  ■ The precision with which actions must comply with specific guidelines
  ■ The amount of authority the leader actually has with team members

Depending on these factors, the manager decides whether to take the autocratic, democratic, participative, or goal-oriented approach. Under different circumstances, the same leader would apply a different leadership style. Detractors reject situational leadership as an attempt to apply an approach based on short-term concerns instead of on the solution of long-term problems.

Best Leadership Style in a Team Setting

The most appropriate leadership style in a team setting might be called *participative leadership taken to a higher level.* Whereas participative leadership in the traditional sense involves soliciting employee input, in a team setting it involves soliciting input from empowered stakeholders, listening to the input, and acting on it. The key difference between traditional participative leadership and participative leadership from a team perspective is that with the latter, the stakeholders providing input are empowered to take the initiative in solving problems and making improvements.

Collecting employee input is not new. However, collecting input, logging it in, tracking it, acting on it in an appropriate manner, working with employees to improve weak suggestions rather than simply rejecting them,

> ### Strategic Planning Tip
>
> **Key Point**
>
> Many people confuse popularity with leadership and, in turn, followership. Leadership and popularity are not the same thing. Long-term followership grows out of respect, not popularity. Good leaders may be popular, but they must be respected.

and rewarding employees for improvements that result from their input extend beyond the traditional approach to participative leadership.

Establishing Followership

People can be team leaders only if those they hope to lead follow them willingly and steadfastly. Followership must be built and, once built, maintained. This section is devoted to a discussion of how team leaders build and maintain followership among the people they hope to lead.

- *Popularity and the leader.*   Many people confuse popularity with leadership and, in turn, followership. Leadership and popularity are not the same thing. Long-term followership grows out of respect, not popularity. Good leaders *may* be popular, but they *must* be respected. Not all good leaders are popular, but all are respected. Team leaders occasionally must make unpopular decisions. This is a fact of life, and it explains why leadership positions are sometimes described as being lonely. Making an unpopular decision does not necessarily cause leaders to lose followership, provided they are seen as having solicited a broad base of input and having given serious, objective, and impartial consideration to that input. Correspondingly, leaders who make inappropriate decisions that are popular in the short run may actually lose followership in the long run. If the long-term consequences of a decision turn out to be detrimental to the team, team members will hold the leader responsible, especially if the decision was made without first collecting and considering employee input.

- *Characteristics that build and maintain followership.*   Team leaders build and maintain followership by earning the respect of those they lead. Here are some characteristics of leaders that build respect (see also Figure 3.3).

    - *Sense of purpose.*   Successful team leaders have a strong sense of purpose. They know who they are and what contributions they make to the organization's success.

**Checklist of
Characteristics That Build
and Maintain Followership**

___✓__ Sense of purpose

___✓__ Self-discipline

___✓__ Honesty

___✓__ Credibility

___✓__ Common sense

___✓__ Stamina

___✓__ Commitment

___✓__ Steadfastness

FIGURE 3.3   Team leaders establish followership by displaying these characteristics.

- *Self-discipline.* Successful team leaders develop discipline. Self-discipline allows team leaders to avoid negative self-indulgence, inappropriate displays of emotion (such as anger), and counterproductive responses to the everyday pressures of the job. Through self-discipline, team leaders set an example of handling problems and pressures with equilibrium and a positive attitude.

- *Honesty.* Successful team leaders are trusted by their followers. They are open, honest, and forthright with team members and with themselves. Team members depend on these leaders to make difficult decisions in unpleasant situations.

- *Credibility.* Successful team leaders have credibility, established by being knowledgeable, consistent, fair, and impartial in all human interactions; setting a positive example; and adhering to the same standards of performance and behavior expected of others.

- *Common sense.* Successful team leaders know what is important in a given situation and what is not. They know that tact is important when dealing with people. They know when to be flexible and when to be firm.

- *Stamina.* Successful team leaders must have stamina. Frequently, they need to be the first to arrive and the last to leave. Their hours are likely to be longer and the pressures they face more intense than those of others. Energy, endurance, and good health are important to those who lead.

- *Commitment.* Successful team leaders are committed to the goals of the team, the people they work with, and their own ongoing personal and professional development. They are willing to do everything within the limits of the law, professional ethics, and company policy to help their team succeed.

- *Steadfastness.*   Successful team leaders are steadfast and resolute. People will not follow a person they perceive to be wishy-washy and noncommittal, or whose resolve they question. Successful team leaders must have the steadfastness to stay the course even when it is difficult.

- *Pitfalls of team leadership.*   The previous section set forth positive characteristics that help team leaders build and maintain the respect and, in turn, the followership of those they hope to lead. Team leaders should also be aware of several common pitfalls that can undermine their followership and the respect they work so hard to earn.

  - *Trying to be a buddy.*   Positive relations and good rapport are important, but team leaders are not the buddies of those they lead. The nature of the relationship does not allow it.

  - *Having an intimate relationship with a team member.*   This practice is both unwise and unethical. A positive team leader–team member relationship cannot exist in this circumstance. Few people can succeed at being both the lover and the leader, and few things can damage the morale of a team as quickly and completely.

  - *Trying to keep things the same when leading former peers.*   The team leader–team member relationship, no matter how positive, is different from the peer–peer relationship. This can be a difficult adjustment to make, but it must be made if the team leader is to succeed.

Strategic Planning Tip

Key Point

Ethical behavior is that which falls within the limits prescribed by morality. Doing what is ethical is often called "doing the right thing."

Ethics in Team Leadership

There are many definitions of the term *ethics*, and no one definition has been universally accepted. The often conflicting and contradictory interests of workers, customers, competitors, and the general public result in a propensity for ethical dilemmas in the workplace. Ethical dilemmas in the workplace are often more complex than other ethical situations. They involve social expectations, competition, and responsibility as well as the potential consequences of choices and decisions to the organization's many constituencies.

In the discussion of ethics, the terms *conscience, morality* and *legality* are frequently heard. Although these terms are closely associated with ethics, they do not by themselves define it. For the purposes of this book, ethics is defined as follows:

> **Ethics** is the application of morality. **Ethical behavior** means "doing the right thing" within a moral framework.

Morality refers to the values that are widely subscribed to and fostered by society in general and individuals within society. The field of ethics attempts to apply reason in determining rules of human conduct that translate morality into everyday behavior. Ethical behavior is that which falls within the limits prescribed by morality. Doing what is ethical is often called "doing the right thing."

How, then, does a team leader know if someone's behavior is ethical? Ethical questions are rarely black and white. They typically fall into a gray area between the two extremes of clearly right and clearly wrong; this gray area is often further clouded by personal experience, self-interest, point of view, and external pressure.

- *Ethical guidelines.* Before presenting guidelines team leaders can use in sorting out matters that fall into the gray area between clearly right and clearly wrong, it is necessary to distinguish between *legal* and *ethical.* They are not always synonymous.

  It is not uncommon for people caught in the practice of questionable behavior to use the "I didn't do anything illegal" defense. Behavior can be well within the scope of the law and still be unethical. The following guidelines for determining ethical behavior assume that the behavior is legal:

  - Apply the *morning-after test.* If you make this choice, how will you feel about it and yourself tomorrow morning?
  - Apply the *front-page test.* If it was printed as a story on the front page of your local newspaper, how would you feel about your decision?
  - Apply the *mirror test.* If you make this decision, how will you feel about yourself when you look in the mirror?
  - Apply the *role-reversal test.* Mentally trade places with the people affected by your decision. How does it look through their eyes?
  - Apply the *commonsense test.* Listen to what your instincts tell you. If it feels wrong, it probably is.

- *Team leader's role in ethics.*   Using the guidelines set forth in the previous section, team leaders should be able to make responsible decisions concerning ethical choices. Unfortunately, deciding what is ethical is often easier than *doing* what is ethical. In this regard, trying to practice ethics is like trying to diet. It is not just a matter of knowing you should cut down on your caloric intake, it is also a matter of following through and actually doing it.

  It is this fact that defines the role of team leaders with regard to ethics. Team leaders have a three-part role. First, they are responsible for setting an example of ethical behavior. Second, they are responsible for helping the team make the right decisions when facing ethical questions. Third, team leaders are responsible for helping team members follow through on the ethical option once the appropriate choice has been identified. In carrying out these responsibilities, team leaders can adopt one of the following approaches:

  - Best-ratio approach
  - Black-and-white approach
  - Full-potential approach
    - *Best-ratio approach.*   The best ratio approach is the pragmatic option. Its philosophy is that people are basically good, and under the right circumstances they will behave ethically. However, under certain conditions they can be driven to unethical behavior. Therefore, the team leader should create conditions that promote ethical behavior and maintain the best possible ratio of good choices to bad. When hard decisions must be made, team leaders should make the choice that will do the most good for the most people. This is sometimes referred to as *situational ethics.*
    - *Black-and-white approach.*   In the black-and-white approach, right is right and wrong is wrong. Circumstances and conditions are irrelevant. The team leader's job is to make ethical decisions, carry them out, and help employees choose the ethical route regardless of circumstances. When difficult decisions must be made, leaders should make fair and impartial choices and deal with the consequences.
    - *Full-potential approach.*   Under the full-potential approach, team leaders make decisions based on how they will affect the organization's ability to achieve its full potential. The underlying philosophy is that people are responsible for helping their organizations realize their full potential within the confines of morality. Choices that achieve this goal without infringing on the rights of others are considered ethical.

Summary

1. Strategic planning is important to organizations for five major reasons: (1) people need direction—they want to know where their organization is going (direction rationale); (2) people want to take control over their lives and try to shape the future (future-shaping rationale); (3) people need to know where they fit into the big picture (big-picture rationale); (4) people need a beacon in the distance to guide them through the everyday storms and fog of work (beacon analogy); and (5) people need to feel that they are a part of deciding where their organization is going and how it will get there (stakeholder buy-in).

2. Stakeholders are more likely to help carry out a plan if they buy into it. Buy-in means more than just accepting the strategic plan, it means believing in it and wanting to make it a reality. Being involved in a meaningful way in the process of planning helps stakeholders buy into the finished plan. Buy-in results in the following benefits to an organization: sense of purpose, enthusiasm, commitment, control, less stress, focus, and motivation.

3. The strategic planning team should consist of individuals who meet one or more of the following criteria: (a) have the authority to commit the organization to a given direction; (b) have the ability to think creatively; (c) have the self-confidence to interact with executives without being intimidated; and (d) have the ability to represent a specific constituency.

4. The size of the strategic planning team will vary according to the size and complexity of the organization. A good rule of thumb is to establish a team with enough members to ensure that all constituencies in the organization are represented, but not so many members that the team is too cumbersome to manage. Most organizations can pull together an effective team with six to ten members. Do not be concerned about having an even number of members. Strategic planning teams are not democracies. They don't vote. Rather, they work for consensus.

5. You can broaden the involvement of stakeholders beyond just the planning team. Being involved in the process and being on the team are not mutually exclusive concepts. An organization can involve many stakeholders in the process without actually naming them to the planning team. This is done by establishing ad hoc committees to complete specific tasks relating to the development of the plan. Each committee reports to one of the members of the planning team.

6. Certain abilities are necessary for effective team leadership. These include the ability to: (a) overcome resistance; (b) broker the needs of constituent groups; and (c) establish an ethical framework. The best leadership style in a team setting is participative leadership taken to a higher level. Good leaders can maintain followership by exemplifying such characteristics

as a sense of purpose, self-discipline, honesty, credibility, common sense, stamina, commitment, and steadfastness.

Key Terms and Concepts

Autocratic leadership	Future-shaping rationale
Balanced commitment	Goal-oriented leadership
Beacon analogy	Good communication skills
Best-ratio approach	Honesty
Big-picture rationale	Motivation
Black-and-white approach	Participative leadership
Commitment	Persuasiveness
Common sense	Positive influence
Control	Positive role model
Credibility	Self-discipline
Democratic leadership	Sense of purpose
Direction rationale	Situational leadership
Enthusiasm	Stakeholder buy-in
Ethical behavior	Stamina
Ethics	Steadfastness
Focus	Stress
Full-potential approach	Team membership

Review Questions

1. Explain the direction rationale for strategic planning.
2. What is the future-shaping rationale for strategic planning?
3. Describe the big-picture rationale for strategic planning.
4. Explain the beacon analogy as it relates to strategic planning.
5. How does stakeholder buy-in figure into the strategic planning process?
6. What are the benefits that can accrue to an organization that ensures buy-in when developing a strategic plan?
7. Who should be on the strategic planning team?
8. How can an organization broaden participation in the strategic planning process without having too large a planning team?
9. Explain the importance of leadership to the strategic planning team.
10. What are the characteristics of a good team leader?

11. List and explain the various styles of team leadership.

12. What is the best leadership style in a team setting and why?

13. List and explain the characteristics that build and maintain followership.

14. What is meant by the term *ethics*, and how does the concept relate to team leadership?

15. List and explain three approaches for dealing with ethical responsibilities.

SIMULATION CASES FOR DISCUSSION

The following simulation cases are provided to generate additional thought and discussion about the principles explained in this chapter. Readers are encouraged to consider how the situations presented in these cases might apply to them and to discuss the cases with others interested in strategic planning and execution.

CASE 3.1 Who Needs a Written Plan?

John Parks and his sister Anne started Parks Technical Services, Inc. (PTS) ten years ago. Since that time the company has grown steadily and now employs more than 300 personnel. John is president of the company, and Anne is vice president for marketing. PTS has no strategic plan. John and Anne know where they want to take the company, but this knowledge is in their heads. It is doubtful that anyone else in the company could articulate a corporate vision or mission. Anne has just told her brother that PTS has grown to the point that it needs a written strategic plan. John disagrees. "Who needs a written plan," said John. It was a statement, not a question. "I know where we are going and you know where we are going. I don't see why we need to write any of this down."

Discussion Questions

1. Does a company such as PTS that is enjoying success really need a strategic plan?

2. Join the debate between John and Anne Parks. Does the plan really need to be written down? Why or why not?

CASE 3.2 Why Is Buy-In Important?

John and Anne Parks continue to debate the issue of developing a written strategic plan for PTS. John doesn't like the idea, but Anne is adamant that a written plan be developed. Anne is explaining her view concerning why they need not only to develop a written plan, but to involve a broad base of employees in the process. She has just told John that employee buy-in is critical. "Why is buy-in so important?" asked John. "They work for us. If we develop the plan, it's their job to carry it out. Who cares if they buy into it or not?"

Discussion Questions

1. Choose a side here. Who is right and why?
2. If you were Anne Parks, how would you explain the importance of buy-in to John?

CASE 3.3 Who Should Be on the Planning Team and How Many?

Anne Parks has finally convinced her brother John that PTS needs a written strategic plan. They are now discussing the size and composition of the team. "Why don't we just have the senior executives develop the plan?" asked John. Anne doesn't like this idea. She wants to have broad-based input from across the company. "Well then you tell me, Anne," said John Parks. "Who should be on the team and how many team members should we have?"

Discussion Questions

1. Do you agree with John Parks that the senior executive team should develop the strategic plan? Explain your reasoning.
2. Assume that PTS is organized in the same way as the typical technology company with 300-plus employees. Describe who should be on the strategic planning team and how many members the team should have.

Endnote

1. Bennis, W., and B. Nanus, *Leaders: The Strategies for Taking Charge* (Harper & Row, New York: 1985), 184–186.

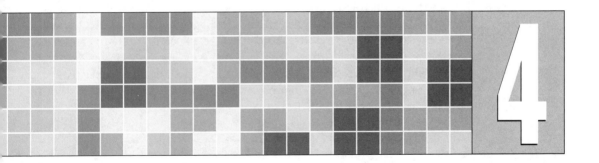

Understand the Strategic Planning Process

"Make no little plans; they have no magic to stir men's blood . . . make big plans, aim high in hope and work."[1]

—J. Vitullo-Martin and R. J. Maskin

OBJECTIVES

- Explain the data collection, analysis, and organization component of the strategic planning process.

- Explain the prediction component of the strategic planning process.

- Describe how to establish direction and formulate strategies as part of the strategic planning process.

- Provide an overview of how to write the strategic plan.

- Provide an overview of how to distribute, communicate, and execute the strategic plan.

- Explain the monitoring, measuring, and adjusting component of the strategic planning process.

The **strategic planning process** is a systematic approach organizations use to create the context within which they will operate. This context shows where the organization is going, how it will get there, and what is required to have a successful journey. This chapter provides an overview of the process the author recommends for strategic planning for competitive advantage. Successive chapters then provide more detailed information about each step in the process.

Strategic Planning Tip

Key Point

Planning is by its nature a future-oriented enterprise. Those who develop plans attempt to look into the future and determine as best they can what the future holds.

OVERVIEW OF THE STRATEGIC PLANNING PROCESS

Many different models of the strategic planning process are available to technology companies. Although the names given to each step in the process differ from model to model, the concepts that make up the various models are essentially the same. Figure 4.1 presents the model the author has found to be the most effective approach for technology companies. This model con-

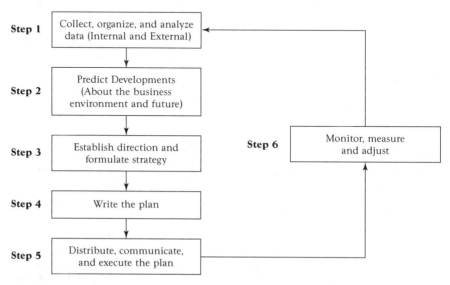

FIGURE 4.1 Overview of the strategic planning process.

sists of six steps or components: (1) collect, organize, and analyze data about your company, the competition, and the business environment; (2) make informed predictions about developments in the business environment and the future; (3) establish the direction for your company and formulate strategy; (4) convert the information from Steps 1 through 3 into a comprehensive, written plan; (5) distribute the plan to all stakeholders, communicate its meaning, and execute; and (6) monitor progress in the execution of the plan, measure actual performance against planned performance, and make adjustments as necessary.

STEP 1: COLLECT, ORGANIZE, AND ANALYZE DATA

The process begins with the strategic planning team (Chapter 3) collecting both internal and external data, analyzing that data, and organizing it in a way that makes it practical and useful to the process. Data collection focuses on several critical questions: (1) What is our financial condition? (2) What is our market? (3) Who is the competition? (4) What is the business environment? (5) What are our strengths and weaknesses? (6) What are our unique value-producing competencies (core competencies)? Step 1 in the process is covered in detail in Chapter 5.

Strategic Planning Tip

Key Point

To the extent possible, the strategic planning team's predictions should be based on solid research and reliable data.

STEP 2: MAKE INFORMED PREDICTIONS

Planning is by its nature a future-oriented enterprise. Those who develop plans attempt to look into the future and determine as best they can what the future holds. In the case of strategic planning for competitive advantage, however, you are doing more than just trying to predict the future—you are trying to shape it. Consequently, this step is less about identifying what *will* be and more about deciding what *can* be. To do this, the strategic planning team must make some predictions. It is important to point out that I speak here not of wild guesses or unrealistic predictions. Rather, I speak of *informed* predictions. The distinction is important. To the extent possible, the strategic planning team's predictions should be based on solid research and reliable data.

Strategic Planning Tip

Key Point

Converting rough draft material into a finished document that is well written and cogently presented is a different kind of undertaking. The quality of the writing and the visual attractiveness of the final plan are important.

In this step the planning team attempts to answer such questions as the following: (1) How will the markets for our product(s) behave in the near and far-term future? (2) How will the competition behave in the near and far-term future? (3) How will changes in the business environment affect our company in the near and far-term future, and what will those changes be? (4) What new opportunities for our company will present themselves in the near and far-term future? (5) What potential threats might affect our company in the near and far-term future and how? Step 2 is covered in detail in Chapter 6.

STEP 3: ESTABLISH DIRECTION AND FORMULATE STRATEGY

Now that you have taken a good hard look at yourself, the competition, markets, and the future, it's time to establish an overall direction for the company and formulate strategy. This step involves adopting a strategic focus, adopting a competitive strategy, developing a vision, and coming to grips with the organization's mission. In this step, the strategic planning team answers the following questions: (1) What of value do we provide to customers? (2) What is the market(s) for what we provide of value? (3) How can we outperform the competition in terms of value? (4) What competitive strategy should we adopt? (5) What is our vision for the organization? (6) What is the mission of our organization? (7) What are the broad goals of our organization? (8) What are the goal-specific strategies of our organization? (9) What are the principles that will guide our organization's operation and decision making as we pursue the vision, mission, and goals? Step 3 is covered in detail in Chapter 7.

STEP 4: WRITE THE PLAN

In Step 3, the strategic planning team developed all of the elements needed to actually convert the strategic plan into a comprehensive written document. This step involves converting the material from Step 3 into final written form. Although it is tempting to do so, it is a mistake to combine Steps

---

Strategic Planning Tip

Key Point

In the author's experience, the first thing organizations do after completing a strategic plan is breathe a sigh of relief and put the plan on a shelf. Unfortunately, in many cases the plan—once put there—never comes off the shelf.

---

3 and 4. Answering the questions posed in Step 3 and roughing out the answers is one kind of undertaking. Converting rough draft material into a finished document that is well written and cogently presented is a different kind of undertaking. The quality of the writing and the visual attractiveness of the final plan are important.

As to the quality of the writing, to support the strategic plan and do their part in executing it, stakeholders must first understand it. A well-written plan will be much easier to understand than one that still reads like a rough draft. As to the attractiveness of the plan, remember that first impressions are lasting. An unattractive presentation of even the best concepts will not invite the attention of potential readers. This is why book publishers put so much effort into the covers of books. Consider the analogy of a speech given by an important dignitary. No matter how well reasoned and logical the speaker's message, and no matter how important the speaker is,

---

STRATEGIC PLANNING PROFILE   Building Value into the Plan at Geraghty & Miller, Inc.

The concept of value plays a critical role in determining whether or not a company can gain a competitive advantage in the marketplace. Providing greater value than competitors is the key. To provide superior value, companies must make value part of their strategic plan.

Geraghty & Miller, Inc.

Geraghty & Miller, Inc. is a technology company that provides environmental consulting, engineering, hydrocarbon, and mediation services. The company's principal areas of operation are protection of groundwater resources, mediation of groundwater contamination, and correction of soil contamination. Geraghty & Miller, Inc. attempts to gain competitive advantage by providing "environmental value" for clients.

Source: Jeffrey Abrahams, The Mission Statement Book (Ten Speed Press, Berkeley, CA: 1999).

if the speech is presented in a dull, monotonous, humdrum manner, people will tune out without giving the message a chance to get through.

Think of the writing and presentation quality of the strategic plan in this way: If the plan were to be the only evidence a stakeholder or potential customer had to make a judgment about your company, what kind of impression would it make? The better the quality of the writing and presentation, the more likely it is that stakeholders will take the time to read the content. Step 4 is covered in detail in Chapter 8.

STEP 5: DISTRIBUTE, COMMUNICATE, AND EXECUTE THE PLAN

In the author's experience, the first thing organizations do after completing a strategic plan is breathe a sigh of relief and put the plan on a shelf. Unfortunately, in many cases the plan—once put there—never comes off the shelf. A strategic plan is a management tool. Like any other tool, it must be taken off the shelf and used in order to do any good.

As soon as the strategic plan has been completed, it should be broadly distributed throughout the organization. Every stakeholder should either receive a copy or have easy access to one. In addition to distributing the plan, the strategic planning team should do what is necessary to communicate its intent and meaning. Never assume that just distributing the plan means it will be read and understood. Rather, be proactive in providing small group sessions in which designated management personnel explain the plan, communicating not just its contents, but also its meaning.

Once the strategic plan has been widely distributed and effectively communicated throughout the organization, implementation should begin immediately. This is the execution phase of the process. It involves conducting a series of roadblock analysis sessions to identify factors that might inhibit execution of the plan and developing ways to overcome or avoid these factors. It also involves delegating, making specific assignments, establishing schedules, setting deadlines, and reinforcing progress. Step 5 is covered in detail in Chapter 9.

Strategic Planning Tip

Key Point

Like any plan, much of a strategic plan is premised on predictions and assumptions about the future. Even with the best data available and even with the most thorough analysis of that data, some aspects of the plan will prove unrealistic or off base during the execution phase.

STEP 6: MONITOR, MEASURE, AND ADJUST

A strategic plan is just that—a plan. Like any plan, much of a strategic plan is premised on predictions and assumptions about the future. Even with the best data available and even with the most thorough analysis of that data, some aspects of the plan will prove unrealistic or off base during the execution phase. Consequently, it is important to monitor results, measure expected performance, and—when necessary—make adjustments to the plan. Step 6 is covered in detail in Chapter 10.

Figure 4.2 is a checklist that can be used as a quick-reference tool for guiding your organization through the strategic planning process. This checklist can be invaluable in keeping members of the strategic planning team focused and on track. However, it should not be substituted for the more in-depth material presented in the remaining chapters of this book.

Checklist
STRATEGIC PLANNING PROCESS

Step 1: Collect, Organize, and Analyze Data

Use the data collected to answer the following questions:

1. What is our organization's financial condition?

2. What is our market?

3. Who is the competition?

4. What is the business environment?

5. What are our strengths and weaknesses?

6. What are our unique value-producing competencies?

Step 2: Make Informed Predictions

1. How will the markets for our product(s) behave in the near and far-term future?

2. How will the competition behave in the near and far-term future?

3. How will changes in the business environment affect our company in the near and far-term future and what will those changes be?

4. What new opportunities for our company will present themselves in the near and far-term future?

5. What potential threats might affect our company in the near and far-term?

FIGURE 4.2   The six step process to successful strategic planning.

Step 3: Establish Direction and Formulate Strategy

1. What of value do we provide to customers?
2. What is the market(s) for what we provide of value?
3. How can we outperform the competition in terms of value?
4. What competitive strategy should we adopt?
5. What is our vision for the organization?
6. What is the mission of our organization?
7. What are the broad goals of our organization?
8. What are the goal-specific strategies of our organization?
9. What are the principles that will guide our organization's operation and decision-making as we pursue the vision, mission, and goals?

Step 4: Write the Plan

Convert all of the information from the previous steps into a well-written, attractive document. Make sure that your organization's strategic plan has at least the following elements:

1. Competitive strategy statement (this element may be left off if it is apparent in either the vision or the mission statements).
2. Vision statement.
3. Mission statement.
4. Guiding principles.
5. Broad strategic goals.
6. Goal-specific strategies.

Step 5: Distribute, Communicate, and Execute the Plan

1. Produce a sufficient number of copies of the finished plan to accommodate a broad-based distribution among stakeholders.
2. Distribute the plan widely among stakeholders.
3. Conduct numerous small-group sessions to explain the contents of the plan and the plan's meaning to stakeholders.
4. Conduct a series of roadblock analysis sessions to identify factors that might inhibit implementation of the plan.
5. Delegate responsibility for implementation of all aspects of the plan, make specific assignments, establish schedules, and set deadlines.

Step 6: Monitor, Measure, and Adjust

1. Monitor all assignments, schedules, and deadlines continually.
2. Measure actual performance against planned performance.
3. Adjust the plan and all activities relating to it as necessary based on what is learned while monitoring and measuring.

FIGURE 4.2 Continued

Summary

1. The strategic planning process consists of six steps: (1) collect, organize, and analyze data; (2) make informed predictions; (3) establish direction and formulate strategy; (4) write the plan; (5) distribute, communicate, and execute the plan; and (6) monitor, measure, and adjust.

2. Step 1 of the strategic planning process—collect, organize, and analyze data—involves answering the following questions: What is our financial condition? What is our market? Who is the competition? What is the business environment? What are our strengths and weaknesses? What are our unique value-producing competencies?

3. Step 2 of the strategic planning process—make informed predictions—involves answering the following questions: How will our markets behave in the future? How will the competition behave in the future? How will changes in the business environment affect our company, and what will those changes be? What new opportunities will present themselves in the future? What potential threats might affect our company in the future?

4. Step 3 of the strategic planning process—establish direction and formulate strategy—involves answering the following questions: What of value do we provide to customers? What is the market(s) for what we provide of value? How can we outperform the competition in terms of value? What competitive strategy should we adopt? What is our vision? What is our mission? What are our broad goals? What are our goal-specific strategies? What are our guiding principles?

5. Step 4 of the strategic planning process—write the plan—involves converting all of the information from Steps 1–3 into a well-written, attractive document that has the following elements: competitive strategy statement, vision statement, mission statement, guiding principles, broad strategic goals, and goal-specific strategies.

6. Step 5 of the strategic planning process—distribute, communicate, and execute—involves taking the following action: producing a sufficient number of copies of the finished plan to accommodate a broad-based distribution among stakeholders; distributing the plan widely among stakeholders; conducting numerous smallgroup sessions to communicate the contents and meaning of the plan; conducting a series of roadblock analysis sessions; delegating responsibility for implementation of all aspects of the plan; making specific assignments; establishing schedules; and setting deadlines.

7. Step 6 of the strategic planning process—monitor, measure, and adjust—involves taking the following action: monitoring all assignments, schedules, and deadlines continually; measuring actual performance against planned performance; and making any necessary adjustments.

Key Terms and Concepts

Collect, organize, and analyze data

Distribute, communicate, and execute the plan

Establish direction and formulate strategy

Make informed predictions

Monitor, measure, and adjust

Strategic planning process

Write the plan

Review Questions

1. Explain all of the various elements of the data collection, organization, and analysis step of the strategic planning process.
2. Explain all of the various elements of the informed prediction step of the strategic planning process.
3. Describe how to establish direction and formulate strategies as part of the overall strategic planning process.
4. Provide an overview of how to write the strategic plan.
5. Summarize an effective approach for distributing, communicating, and executing the strategic plan.
6. Explain the monitoring, measuring, and adjusting component of the strategic planning process.

SIMULATION CASES FOR DISCUSSION

The following simulation cases are provided to generate additional thought and discussion about the principles explained in this chapter. Readers are encouraged to consider how the situations presented in these cases might apply to them and to discuss the cases with others interested in strategic planning and execution.

CASE 4.1 I'm Not Comfortable Basing So Much on Predictions

"It seems to me that we are betting the future of our company on guesswork," said Andrew Buzland. "I'm not comfortable basing so much on predictions."

Buzland is the CEO of Defense Technologies, Inc. (DTI). He has decided that his company needs to develop a comprehensive strategic plan. However, now that the process is underway, he is growing increasingly uncomfortable with it.

"Review this step for me again," said Buzland to his planning team.

Discussion Questions

1. What is the key term being left out of this discussion about predictions at DTI?
2. How would you respond to Buzland if you were part of DTI's strategic planning team?

CASE 4.2 We've Already Written the Plan—Why Rewrite it?

The DTI strategic planning team has now completed Steps 1–3 in the process. While completing these steps, they roughed out a good solid draft of a plan. One member of the team thinks the draft is sufficient. He says: "We've already written the plan—why rewrite it?"

Discussion Questions

1. Who would you say the person asking the question in this case views as the audience for DTI's strategic plan?
2. How would you answer the question posed in this case?

CASE 4.3 Now That the Plan Is Done, Maybe We Can Get Back to Work

Andrew Buzland, CEO of DTI, chaired the strategic planning process that just resulted in a completed plan for the company. However, the CEO has not been comfortable with the process from the beginning. He appears to think it has taken him and the other members of the strategic planning team away from their "real jobs." As Buzland flipped through the first copy of the plan off the press, he commented to his secretary, "Now that the plan is done, maybe we can get back to work." He then placed his copy on a shelf in his office and turned to his computer to check his e-mail.

Discussion Questions

1. Based on Buzland's attitude toward the process and the completed plan, what do you think will come of the strategic planning process at DTI?
2. What should the CEO have done when the finished copies of the new plan came off the press?

Endnote

1. Vitullo-Martin, J., and R. J. Maskin, *Executive's Book of Quotations* (Oxford University Press, Oxford, England: 1994), 217.

Collect, Organize, and Analyze Data

When you are drowning in numbers you need a system to separate the wheat from the chaff.[1]

—Anthony Adams,
Campbell Soup Company

OBJECTIVES

- Explain how to assess the financial condition of an organization.

- Explain how to assess the strengths and weaknesses of an organization.

- Describe how to identify organizational competencies that produce value.

- Summarize how to conduct a market segmentation analysis for an organization.

- Summarize how to analyze an organization's competition.

A good strategic plan is based on a combination of facts and informed forecasts. The facts, if properly collected, organized, and analyzed, will paint an accurate picture of what currently exists. Informed forecasts help paint a picture of the future. These forecasts are a combination of what will be and what can be. Remember, one of the reasons for developing a strategic plan is to shape the future to your organization's benefit.

The step in the strategic planning process covered in this chapter involves collecting, organizing, and analyzing facts about your company, markets, and the competition. This might be the most difficult step in the strategic planning process because it requires an organization to stand in front of a mirror and take a good hard look at itself. For some executives accustomed to showing the world their company's best possible face, this can be like stripping off their clothes and standing naked before the world. Nevertheless, this step must be completed in a manner that is straightforward, open, objective, and honest. Covering up unpleasant findings amounts to hiding information that might help the organization improve and compete. It would be like going to the doctor for a checkup, finding a cancerous growth, and hiding or ignoring the fact. The quality of the data collected in this step will either validate or invalidate the remainder of the planning process.

Strategic Planning Tip

Key Point

Part of the strategic planning process is about deciding what is possible. This decision cannot be made without a full and accurate assessment of your organization's financial condition.

ASSESSING YOUR ORGANIZATION'S FINANCIAL CONDITION

Assume you have outgrown your current home and want to build a bigger one. The prudent thing to do before meeting with an architect to develop plans would be to assess your financial condition. Before developing the plans for your new house, you would want to know how much house you can afford. You would want to compare your assets and liabilities, consider your current income, and project your future income. You might also analyze the various sources of your income to determine how reliable they are and which might fluctuate unpredictably. With this financial assessment completed, you would know which home plans are realistic.

The same types of financial assessments should be made by organizations beginning the strategic planning process. Part of the strategic planning process is about deciding what is possible. This decision cannot be made without a full and accurate assessment of your organization's financial condition. In assessing your organization's financial condition, at least three of the most common financial monitoring reports should be examined: statement of assets and liabilities, operating budgets, and the profit analysis statement.

Assets and Liabilities

The strategic planning team begins the financial assessment by asking the organization's finance and accounting department to prepare a comprehensive balance sheet showing all assets, all liabilities, and the organization's net worth. A comprehensive balance sheet will begin to paint an accurate picture of your organization's financial condition. It will become especially important should the organization eventually adopt a strategic plan that will result in additional debt.

Operating Budgets

The strategic planning team should ask the finance and accounting department to prepare a comprehensive summary of the organization's operating budgets for the current year and for at least the last three years. Depending on the age of the organization, you might want to go back as far as ten years. The further back you go, the more accurate the analysis will be. Your analysis amounts to comparing budgeted performance against actual performance and identifying any trends that might indicate negative patterns or issues of concern. The ideal situation is to find that the organization consistently operates within its budget and accurately forecasts when developing budgets.

Profit Analysis

The strategic planning team should ask the finance and accounting department to prepare profit analysis statements that answer such questions as: (1) Where do our profits come from? (2) Which of our products are most profitable? (3) Which of our divisions, branches, or plants are most profitable (if applicable)? (4) What are the margins for each of our products or profit centers? (5) Which of our markets are the most profitable? To answer these types of questions, the strategic planning team will need profitability analysis statements prepared from several different perspectives (e.g., product, market segment, customer demographics).

With comprehensive, well-organized financial information in hand, the strategic planning team can make an informed determination about which

Strategic Planning Tip

Key Point

People in an organization often feel threatened when asked to admit to weaknesses in themselves, their functional units, or their direct reports. People who feel threatened might try to cover up weaknesses out of fear of being blamed for them or out of a sense of protective loyalty to their team members.

products, markets, and customers generate the bulk of the organization's profits. This information will be especially important later in the process when the strategic planning team attempts to forecast the organization's future profitability.

ASSESSING YOUR ORGANIZATION'S STRENGTHS AND WEAKNESSES

Most strategic planning models include what is typically called a **S.W.O.T. analysis,** an in-depth study of the organization's strengths, weaknesses, opportunities, and threats. In the model presented in this book, the strengths and weaknesses portion of the S.W.O.T. analysis is undertaken at this point. This can be a difficult step in the data collection process because of a natural human tendency to accentuate strengths and minimize weaknesses.

This tendency is caused by the **threat factor.** People in an organization often feel threatened when asked to admit to weaknesses in themselves, their functional units, or their direct reports. People who feel threatened might try to cover up weaknesses out of fear of being blamed for them or out of a sense of protective loyalty to their team members.

Strategic Planning Tip

Key Point

Before beginning the strengths-and-weaknesses assessment, the strategic planning team should take the time to prepare the organization. The organization's CEO and senior executives should play leadership roles in carrying out this task.

Another problem that can inhibit the assessment of organizational strengths and weaknesses is the methodology—how you go about conducting the assessment. The problems associated with securing accurate, objective information when assessing organizational strengths and weaknesses can be overcome by thorough preparation and the selection of an effective methodology. On the other hand, poor preparation and an ineffective methodology will just compound the data collection problems.

Preparing the Organization for a Strengths-and-Weaknesses Assessment

Before beginning the strengths-and-weaknesses assessment, the strategic planning team should take the time to prepare the organization. The CEO and senior executives should play leadership roles in carrying out this task. An effective approach is to conduct a series of small group meetings led by the CEO or a senior executive in which the purpose and importance of the assessment are explained and discussed.

Since frank give-and-take dialogues are critical in these small group meetings, it is best to organize participants according to personnel classifications. Have managers come together in one meeting, first-line supervisors in another, and employees in another series of meetings. How many meetings will be necessary will, of course, depend on the size of the company and number of operational locations. Participants will be less reluctant to speak out when in the presence of colleagues and peers rather than people of higher authority.

The message that should be given to all participants is the same:

> We need an accurate assessment of what we do well and what we do poorly, and the strategic planning team needs your help in making these determinations. This information will be used for one reason and one reason only—to help us develop a plan for success for our organization. The information will not be used in any way that is punitive, negative, or harmful to those providing it. If we are going to survive and thrive in today's competitive business environment, we must know what we do well and what we need to improve.

Selecting an Effective Assessment Methodology

There are several different methods for assessing an organization's strengths and weaknesses. The three recommended by the author are as follows: (1) functional unit method, (2) value tree method, and (3) key performance measures method.

Valero Energy Corporation converts low-cost feedstocks into high-value, high-quality energy-related products. The company has identified the following strengths that help make it competitive:

- "Producing environmentally clean products in environmentally clean facilities."
- "Aggressively pursuing growth opportunities, both domestically and internationally."
- "Continued development of all employees, our number-one asset."
- "Providing a challenging, rewarding environment that facilitates creative thinking, teamwork, communications."
- "Customer satisfaction by providing reliable and responsive products and services."
- "Taking a leadership role in our communities by providing company support and encouraging involvement."

Source: Valero Energy Corporation Web site (http://www.valero.com)

Functional Unit Method

With this approach, the organization's senior executives assemble the heads of all functional units in their areas of responsibility into fact-finding teams. All functional unit heads are given responsibility for identifying the strengths and weaknesses in their units and for providing a summary report to the appropriate senior executive. Each senior executive, in turn, compiles an overall summary for his various units.

Before functional unit heads begin identifying the strengths and weaknesses in their areas of responsibility, they need a framework that gives structure to their efforts. An effective framework is shown in Figure 5.1. Using this framework as a guide, each functional unit head assesses and summarizes the strengths and weaknesses of the people, processes, technologies, culture, and financial resources of his or her unit. The sum of all such unit-by-unit assessments becomes the organization's list of strengths and weaknesses.

Value Tree Method

With this method, the strategic planning team forms an ad hoc committee of the organization's top managers. Depending on the composition of the strategic planning team, this ad hoc committee and the planning team might be one and the same. However, typically the membership of the strategic planning team includes participants other than the organization's top managers. When this is the case, an ad hoc committee is formed with membership restricted to senior management.

Framework for Analyzing
ORGANIZATION STRENGTHS AND WEAKNESSES

Objectively analyze organizational strengths and weaknesses in your area(s) of responsibility in each of the following areas:

- People
- Processes
- Technology
- Culture
- Financial Resources

FIGURE 5.1   A structured approach will yield better results.

The ad hoc committee uses the value tree (Chapter 2, Figure 2.1) to structure its analysis efforts. All members of the ad hoc committee are responsible for identifying and compiling a summary of the organization's strengths and weaknesses in their areas of responsibility. Organizational strengths and weaknesses are recorded under each component of the value tree, Figure 5.2.

With this method, each member of the ad hoc committee solicits input from his or her various managers and supervisors, adds their observations to his or her own, and compiles a summary of strengths and weaknesses for those areas of responsibility. First-line supervisors are encouraged to solicit input from employees before providing their own.

Strengths and Weaknesses Analysis
VALUETREE METHOD

Objectively identify organizational strengths and weaknesses in the following components of the valuetree:

- Support Functions
- Input Processes
- Conversion Processes
- Output Processes
- Sales Marketing Functions
- Product Issues

FIGURE 5.2   Valuetree framework.

Key Performance Measures Method

With this method, the strategic planning team brainstorms to identify the organization's key performance indicators—those things it must do well to succeed. The key performance indicators, in turn, are used as a framework for conducting the strengths and weaknesses assessment. Key performance indicators can vary widely depending on the type of company, products, and markets in question. Figure 5.3 lists some factors that can affect performance. However, it should be noted that this list is provided only to trigger your thought processes. The list is not comprehensive, but it can be used as a starting point for identifying key performance indicators for your company.

In some cases the strategic planning team might be able to identify all of the key performance indicators relating to a given function, issue, or factor. However, in most cases it will be necessary to solicit input from managers and supervisors using the same approach applied with the value tree method.

Compiling the Strengths Component of the Assessment

Once all strengths for the organization have been identified and compiled, they must then be prioritized. Strengths are prioritized according to their relative importance to the organization's success. For example, an organization might identify the following as its principal strengths:

1. Efficient input processes
2. Effective human resources department
3. Product brand loyalty of customers
4. Excellent customer service
5. Effective marketing/sales function

The strategic planning team must now decide how to prioritize these strengths so as to make the best use of them in gaining advantages over the competition. This task is accomplished by posing the following question: Which of these strengths contributes the most to our company's success?

Using a scale with ratings ranging from 1 to 5 such as the following can be a helpful technique for prioritizing organizational strengths:

- 5 = Critical
- 4 = Very important
- 3 = Important
- 2 = Somewhat important
- 1 = Helpful

Examples of
FACTORS TO CONSIDER
When Assessing
ORGANIZATIONAL STRENGTHS AND WEAKNESSES

- Personnel performance
- Process performance
- Equipment performance
- Systems
- Software
- Tools
- Work procedures/methods
- Facility layout
- Production levels
- Inventory
- Culture
- Customer satisfaction/relations/service
- Marketing/sales performance
- Financial resources/strength
- Product pricing/cost
- New product development
- Time to market for new products
- Time to profit for new products
- Strength of the management team
- Performance of support functions
- Creativity/innovation
- Teamwork
- Change management
- Product branding

FIGURE 5.3   Examples of factors that can affect organizational performance.

The factor given the highest rating is recorded first on the list of priorities, and so on down the list. If two or more factors receive the same rating, list them with their equal numerical rating in parentheses beside each respective entry.

Key Point

The results of the strengths-and-weaknesses assessment are used in several ways: (1) when selecting the organization's overall competitive strategy; (2) when deciding how best to attack the competition; (3) when deciding where to avoid the competition; and (4) when deciding how to most effectively apply resources to strengthen weak areas and improve the organization's ability to compete.

Compiling the Weaknesses Component of the Assessment

Once all weaknesses for the organization have been identified and compiled, they must be prioritized. Weaknesses are prioritized according to their relative impact in harming the organization's ability to compete. For example, an organization might identify the following weaknesses:

1. Time to market for new products (too long)
2. Weak conversion processes
3. Product pricing (not competitive)
4. Production levels

The strategic planning team must prioritize these weaknesses so that it knows either where to avoid the competition or what it must improve in order to compete. Applying a scale ranging from 1 to 5 such as the following can be a helpful technique for prioritizing organizational weaknesses:

- 5 = Extremely harmful
- 4 = Very harmful
- 3 = Harmful
- 2 = Somewhat harmful
- 1 = Marginally harmful

Key Point

The way to win consistently over time in a competitive business environment is to: (1) know what your customers value, and (2) do a better job than the competition of providing what customers value.

Overall List
ORGANIZATIONAL STRENGTHS AND WEAKNESSES
ABC Technologies, Inc.

Strength:

- Product brand loyalty of customers (5)
- Excellent customer service (4)
- Effective marketing/sales (3)
- Efficient input processes (2)
- Effective human resources department (2)

Weaknesses:

- Weak conversion processes (5)
- Production levels (4)
- Time to market for new products (4)
- Product pricing (3)

FIGURE 5.4   Sample list for a hypothetical company.

Using the Results of the Assessment

The results of the strengths-and-weaknesses assessment are used in several ways: (1) when selecting the organization's overall competitive strategy; (2) when deciding how best to attack the competition; (3) when deciding where to avoid the competition; and (4) when deciding how to most effectively apply resources to strengthen weak areas and improve the organization's ability to compete.

Figure 5.4 is an example of an organization's list of strengths and weaknesses in order of priority. The strategic planning team decided that product brand loyalty was the organization's most significant strength. Consequently, when developing the organization's strategic plan, this strength would be exploited to the company's greatest possible benefit. The organization's strategic planners should also find ways to take full advantage of the other strengths listed.

The strategic planning team identified weak conversion processes as the organization's most harmful weakness. This weakness most likely is the cause of the poor production levels and at least a contributing cause to the product-pricing problem. Time to market for new products is also a harmful factor in that it gives the competition too much time to respond or even to beat the organization to market with its own new products. This company will have to strengthen its conversion processes and cut down on the time it takes to bring a new product to market if it hopes to compete over the long run. These facts should be the basis for improvement goals that become part of the strategic plan.

IDENTIFYING CORE COMPETENCIES THAT PRODUCE VALUE

The way to win consistently over time in a competitive business environment is to: (1) know what your customers value and (2) do a better job than the competition of providing what customers value. Your organization's competitive position vis-à-vis the competition is a direct result of the effective application of core competencies that produce value.

A core **competency that creates value** is a mix of human talents and attitudes effectively combined with organizational processes and technologies in ways that, when taken together, allow your organization to provide superior value to customers. In the previous section, methods for identifying organizational strengths were presented. It is important to note here that an organizational strength is not necessarily a core competency that produces value. An organizational strength can be a core value-producing competency, but it might not be. On the other hand, a core value-producing competency is by definition an organizational strength.

Characteristics of Core Competencies That Produce Value

A core competency that produces value is a capability that separates your organization from the competition—a differentiator. Think of a major league baseball team. If the team's fans value consistent winning or championship seasons, a core value-producing competency might be an outstanding pitching staff. Another such competency might be a strong lineup of power hitters. This team's hitters and pitchers represent core value-producing competencies because they have the following characteristics:

1. These players are critical to providing what customers value most.
2. The team combines human talent and attitudes with organizational processes and technologies in highly effective ways.
3. These competencies are effective differentiators from a competitive perspective.
4. These competencies are sufficiently unique to be difficult for the competition to match, develop, acquire, or replicate.

What if the team has an especially strong public relations department that produces the most attractive, most informative team programs in the league? This would be considered an organizational strength, but would

---

Strategic Planning Tip

Key Point

To identify core competencies that produce value, you must first know what customers value most about your products and services. At first glance, this might sound like a statement of the obvious, but organizations must be careful here. One of the reasons some organizations fail is they make inaccurate assumptions about what their customers value.

not be considered a core value-producing competency because it has little or nothing to do with providing what fans value most—consistent winning and championship seasons. This same distinction applies when identifying the core competencies that produce value for your organization.

Begin with a Solid Understanding of What Customers Value Most

To identify core competencies that produce value, you must first know what customers value most about your products and services. At first glance, this might sound like a statement of the obvious, but organizations must be careful here. One of the reasons some organizations fail is they make inaccurate assumptions about what their customers value. It is not uncommon to find that customers purchase a given product for a reason or use that differs from the design's intent.

Organizations that secure contracts by responding to bid specifications can discern from the specifications what the customer values. However, organizations that secure their business in more traditional ways don't have this luxury. A good rule of thumb for such businesses is: *Never assume you know what customers value most—ask them.*

Strategic Planning Tip

Key Point

An effective way to identify an organization's core value-producing competencies is to apply the characteristics of these competencies to the organization's overall set of competencies.

Methods for
IDENTIFYING WHAT CUSTOMERS VALUE

- Reviewing warranty records
- Reviewing customer-service records
- Interviewing personnel who have the most frequent contact with customers
- Conducting customer interviews
- Conducting customer focus groups
- Soliciting written customer feedback
- Recording, compiling, and reviewing customer complaints

FIGURE 5.5 Do not assume you know what customers value—look, listen, and ask.

Consider the example of fast-food restaurants such as McDonald's. In the early years of the concept, had fast-food pioneers assumed that the taste of a hamburger was what their customers would value most, they never would have sold their first franchise. Early fast-food planners knew (or at least suspected) that societal changes were going to produce a new type of customer who valued convenience, low prices, and fast service above all else. With all of the menu and other changes the various fast-food restaurants have gone through over the years, what their customers value most has not changed; it's still convenience, low prices, and fast service. To stay ahead, managers in fast-food chains must know what core competencies produce these highly valued characteristics in their restaurants.

Rather than assume they know what customers value most, organizations are wise to look, listen, and ask. Figure 5.5 lists methods organizations can use to determine what customers value most about their products and services. It is not necessary to use every method on the list, but it is wise to find ways to solicit customer input and to encourage customer feedback.

Strategic Planning Tip

Key Point

It is not uncommon for a strategic planning team to find that the organization has no core value-producing competencies. It has competencies, but they do not rise to the core value-producing level.

Apply the Characteristics of Core Value-Producing Competencies as Criteria

An effective way to identify an organization's core competencies that produce value is to apply the characteristics of these competencies to the organization's overall set of competencies. The strategic planning team begins with a list of what the organization's customers value most about its products and services. It then poses the following question: What competencies do we have or do we need to have to outperform the competition in terms of what our customers value most? With input from the appropriate personnel, this question can be quickly answered.

In answering this question, the strategic planning team will determine either that core value-producing competencies currently exist in the company or that they do not exist and must be developed. To be considered a core value-producing competency, a given organizational competency must have all of the characteristics explained earlier and restated here:

1. Is critical to providing what customers value most.
2. Combines human talent and attitudes with organizational processes and technologies in highly effective ways.
3. Effectively differentiates your organization's products and services from those of the competition.
4. Is sufficiently unique to be difficult for the competition to match, develop, acquire, or replicate.

Organizational competencies that fail to satisfy all four of these criteria cannot be considered core value-producing competencies. They are competencies and might even be organizational strengths, but they are not core competencies that produce value. For example, an organization might gain a temporary competitive advantage by being the first competitor to purchase a new technology. The effective use of this technology is an organizational strength. However, the technology in question can and probably will be purchased by other competitors who will soon become just as

---

Strategic Planning Tip

Key Point

One of the ironies of strategic planning is that few companies undergo the process until they are fully operational. Consequently, the market analysis component of the strategic planning process, more often than not, involves *refining* rather than *defining* markets.

proficient using it as your organization. Therefore, it would not be considered a core value-producing competency. On the other hand, if your organization developed and patented the technology in question and it satisfied all four criteria listed earlier, it would represent a core value-producing competency.

It is not uncommon for a strategic planning team to find that the organization has no core value-producing competencies. It has competencies, but they do not rise to the core value-producing level because they fail to satisfy one or more of the four necessary criteria. When this is the case, developing the necessary core competencies that produce value is included in the organization's strategic plan as a high priority goal.

REFINING AND DEFINING MARKETS

One of the ironies of strategic planning is that few companies undergo the process until they are fully operational and have facilities, employees, products, competitors, and debt. In the author's experience, it is rare to work with a start-up company in developing a strategic plan. Consequently, the market analysis component of the strategic planning process, more often than not, involves *refining* rather than *defining* markets.

If yours is a start-up company, at this point in the process you will analyze potential markets and then, based on the analysis, define the best markets for your company. However, if yours is a fully operational company, you are already producing products for some markets. If this is the case, refining your markets is the current task. In either case, an executive decision must be made concerning whether your company plans to be just another player in the game or the dominant player.

Surviving Versus Thriving

Current business literature is replete with articles about surviving in an intensely competitive business environment, an environment that is global in scope. The author believes that just surviving is the wrong goal. Certainly, the globalization of competition has made succeeding in business a

---

Strategic Planning Tip

Key Point

Strategic planning should be about more than just surviving. It should be about thriving. Set your sights on more than just making a respectable showing in the marketplace. Go for the championship.

more daunting challenge than it has ever been. But it has also created a situation in which a market segment can be refined to the point where it is very thin and still has enormous potential because its depth is now global in scope.

Consequently, strategic planning should be about more than just surviving. It should be about thriving. Set your sights on more than just making a respectable showing in the marketplace. Go for the championship. Plan to win consistently year after year. In other words, plan to thrive, not just survive. The best way to dominate is to refine your markets to the point that your organization can be the best in the world at providing value to the customers in that thin but deep market segment.

Market Strategy Review

In Chapter 1, the concept of market strategy was explained. To review, the two basic strategies are low cost and differentiation, with additional versions of each. A low-cost market strategy attempts to reach a broad market consisting of customers who value the product's price above other considerations. The economy car was used in Chapter 1 as an example of a low-cost product strategy. A differentiation strategy focuses on a thinner market segment that values certain product attributes more than price. A luxury car is a product developed for a differentiation market. To employ a combination strategy, the organization attempts to provide as much in the way of differentiated attributes as possible while still attempting to hold the line on pricing (Figure 5.6). A mid-range car is a product developed for a combination market.

A low-cost strategy results in low margins. This, in turn, means that profits depend on volume. A differentiation strategy results in higher margins but lower volume. For example, an economy car might have a profit margin that is just in the hundreds of dollars. Consequently, manufacturers of economy cars must sell a lot of them to make a respectable profit. A luxury car, on the other hand, might have a profit margin in the thousands. Consequently, manufacturers can make a respectable profit on a lower sales volume.

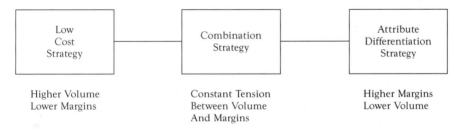

FIGURE 5.6 The choice of market strategy can mean the difference between thriving and just surviving.

With combination strategies, there is always tension between cost and the attributes that differentiate the product. To make an acceptable profit, companies must strike just the right balance between cost and attributes; a difficult challenge at best. What constitutes the right balance is never set in stone. Rather, the necessary balance is a fluid concept that changes continually. A company can strike just the right balance one year and miss it completely the next. In addition, combination strategies generate the most competition. The broad market that is the target of combination strategies has a powerful appeal to companies. But winning consistently in this market requires companies to be best in class at so many different competencies that few can succeed over the long run.

Consequently, the author typically recommends that companies adopt either a low-cost or a differentiation strategy and, of the two, recommends differentiation. Only companies with enormous resources will be able to thrive with a combination strategy. For example, the larger automobile manufacturers of the world typically have products that appeal to low-cost, combination, and differentiation markets. But this approach is unrealistic for most companies. Few companies will succeed in a competitive marketplace trying to be all things to all customers. Rather, the better strategy is to refine your markets to the point that you can be the best in the world at providing superior value in a tightly defined sector. This approach will necessarily mean pursuing market segments that are sliced thinner than usual. But remember that because of globalization, a market segment that, figuratively speaking, is only an inch wide can be a mile deep.

MARKET SEGMENTATION

The ideal market segment is one in which the customers all value the same attributes of your product. When you begin the process of market segmentation, you are looking for market segments in which your organization's most unique competencies will allow your products to stand out from the crowd in ways that are difficult for the competition to match. The more you refine the market segments in question, the fewer competencies at which your organization must excel. This, in turn, allows you to focus your efforts and resources to the point that no competitor can match or replicate them. There are several ways to segment markets. Some of the more widely used are explained in the following paragraphs.

Geographic Segmentation

Some products have natural geographic limitations. For example, manufacturers of ocean-going yachts would not be concerned about the Midwest as a potential geographic market. However, the East, West, and Gulf Coasts of the United States represent potentially lucrative markets. Because of the difficulties and expense associated with shipping large yachts, a company might decide to focus on a certain coastal region and be the dominant manufacturer of yachts in that region.

Customer Demographics Segmentation

Customer demographics must be considered carefully when considering potential markets. For example, a company that produces clothing for teenagers needs a different set of competencies than a company that produces clothing for the elderly. A magazine publisher that produces magazines for bodybuilders needs a different set of competencies than one that produces magazines for sport fisherman. Segmenting markets by customer demographics is one of the most effective ways to refine a broad market into a market segment your organization can dominate.

Purchase Method Segmentation

How do customers actually purchase your product? Do they walk into a store in the local shopping mall? Do they make their purchase over the Internet or from a catalog? Do they solicit bids based on a set of specifications? Segmentation by method of purchase is another way to refine a broader market in a way that might make it easier for your organization to be the dominant player in that market.

Product Type Segmentation

Consider the following examples of products: kitchen stoves, books, and personal computers. Now try to think of all the different versions of these products you can. There are kitchen stoves for residential applications. There are larger and sturdier kitchen stoves for commercial applications such as restaurants. Then there are even larger kitchen stoves for industrial applications such as cook-chill plants that prepare assembly line meals in large quantities for packaging. The broad category of books includes novels, textbooks, how-to manuals, coffee-table books, and many other classifications. Personal computers can be classified as those designed for home and office use, those designed for the factory floor, those designed for military applications, and those designed for numerous other uses. Every one of these different applications or versions in each of these product categories represents a market segment a company could realistically dominate.

Figure 5.7 provides several examples of markets that have been refined by segmentation. These examples are intended to trigger your thought processes as you consider ways to refine segments of your organization's markets in ways that will promote market domination.

ANALYZING MARKET SEGMENTS

Once the strategic planning team has identified what it thinks might be potentially viable market segments, each segment must be carefully and thoroughly analyzed. Figure 5.8 is a guide for analyzing market segments. Few strategic planning teams will be able to answer all of the questions contained in the analysis guide without help. Here, then, is another opportunity to involve additional personnel within the organization, employees who possess specific knowledge and expertise in selected areas.

Involving these subject matter experts in the process will benefit the organization in at least two important ways. First, their involvement will broaden the base of support for the plan once it is completed. Second, their involvement will ensure that the information collected for each market segment will be sufficiently accurate and detailed to allow for a credible analysis.

Using the Market Analysis Guide

The questions set forth in the Guide for Analyzing Market Segments dictate which subject matter experts in the organization will be called on for assistance. Once these personnel have been identified, they are given a copy of the guide and a time frame for collecting the information needed to an-

---

Market Segmentation Examples

Furniture Manufacturing:

- Outdoor and lawn furniture
- Bedroom furniture
- Dining room furniture
- Living room/den furniture
- Office furniture

Engineering Software:

- Civil engineering applications
- Electrical engineering applications
- Mechanical engineering applications
- Aerospace engineering applications

Aircraft Systems Integration Services:

- Military aircraft
- Commercial aircraft
- Corporate aircraft
- Private/personal aircraft

FIGURE 5.7 Refining market segments is an important step toward market dominance.

swer the questions for which they are responsible. It is important to ensure that the information collected for each question is comprehensive enough to provide a clear and complete picture for the strategic planning team.

How Is the Market Segment Defined?

In answering this question, it is important to ensure that the segmentation methodology is either stated or apparent. The answer to this question should also explain critical customer characteristics and buying habits.

What Do Customers in This Market Segment Value Most?

In order to answer this question, it might be necessary to apply some or all of the methods described earlier for identifying customer preferences (e.g., studying warranty records, examining records of customer complaints, interviewing customers, conducting customer focus groups).

Guide for
ANALYZING MARKET SEGMENTS

- How is the market segment defined?
- What do customers in this market segment value most?
- What is the size of the market segment (number of customers and potential sales volume)?
- What is our current share of the market (sales volume and percentage)?
- Who are our competitors in this market segment and what is their market share?
- Are there trends or other significant issues relating to this market segment we should consider?

FIGURE 5.8   Market segments should be analyzed carefully for viability.

What Is the Size of the Market Segment?

The answer to this question should identify the maximum possible sales volume in the market segment in question. This means that before delegating this question to subject matter experts, the strategic planning team must decide how deep the market segment in question will be. Regional? United States? English speaking countries? Global? In addition to the potential sales volume, it is important to identify the number of customers in the market segment.

What Is Our Current Share of This Market Segment?

This question assumes that an existing, operational organization is developing a strategic plan, which is typically the case. Start-up companies may skip this question. Current market share should be recorded in two ways—by percentage and by sales volume.

Strategic Planning Tip

Key Point

One of the worst situations an organization can find itself in is constantly competing for market share with competitors whose products are so similar that customers have no reason to select one product over the other.

Who Are Our Competitors in This Market Segment, and What Is Their Market Share?

All competitors that have even a small share of the market in the segment in question should be identified and listed. The actual market share by both sales volume and percentage should also be listed.

Are There Trends, Issues, or Developments Relating to This Market Segment We Should Consider?

This is a critical question. All of the previous questions might indicate that your organization has an excellent chance of dominating this market segment, but there might be an issue, trend, or development looming just beyond the horizon that could radically alter circumstances. On the other hand, there might also be developments, trends, or issues in the business environment that might potentially work in favor of market dominance for your company. For this reason, it is a good idea to focus as many minds as possible on answering this question.

Strategic Planning Tip

Key Point

Your organization's real competition consists of other companies that pursue the same market segments consisting of customers who value the same things. It is critical to identify these real competitors so that resources can be effectively and efficiently put to use finding ways to outperform them rather than being wasted on strategies to compete with the wrong companies.

ANALYZING THE COMPETITION

One of the worst situations an organization can find itself in is constantly competing for market share with competitors whose products are so similar that customers have no reason to select one product over the other. When this happens, customers typically choose on the basis of price. When price is the only issue, profit margins eventually become so thin that it is impossible to capture enough volume to generate an acceptable profit. This is why the author recommends careful market segmentation and refinement. It is important to be different than the competition in ways that are not price dependent.

One of the goals of market segmentation and refinement is to find ways to permanently, or at least as permanently as possible, differentiate the organization from the competition in ways other than price. To find these ways to differentiate an organization from the competition, you must do more than just analyze your organization's strengths and weaknesses, segment its markets, and carefully analyze each segment. It is also necessary to analyze the competition thoroughly.

When you know the strengths and weaknesses of the competition, your organization can adopt strategies that avoid the strengths and attack the weaknesses. Think of your organization as a pitcher on a professional baseball team. By analyzing the competition, the pitcher learns that the lead-off batter is a good curveball hitter. The pitcher knows he has a weak curveball, so he avoids this batter's strength by throwing fastballs, changeups, and sliders—anything but curves. Correspondingly, if the pitcher knows that a certain hitter has trouble with inside fastballs and one of the pitcher's better pitches is the fastball, this batter is going to see a lot of inside fastballs. The pitcher has adopted strategies for avoiding the competition's strengths and attacking its weaknesses.

Another benefit that can be gained from analyzing the competition is identifying clearly who the real competition is. Not every other organization operating in the same broad markets as yours is necessarily a competitor. Your organization's real competitors are those companies that attract customers for the same or similar reasons as your company. In other words, your organization's real competition consists of other companies that pursue the same market segments consisting of customers who value the same things. It is critical to identify these real competitors so that resources can be effectively and efficiently put to use finding ways to outperform them rather than being wasted on strategies to compete with the wrong companies.

Competition Analysis Guide

When identifying who the competition is, where they are strong, and where they are vulnerable, it is important to ask the right questions. It is also important to avoid wasting valuable time on information that is not sufficiently relevant. Returning to the baseball analogy used earlier, the pitcher should analyze the batters' hitting strengths and weaknesses. Other facts about the batter such as whether he has a strong throwing arm, is good at making double plays, or has trouble catching high fly balls are all interesting, but they are not germane to the issue of hitting. The competitive issue for the pitcher is hitting skills.

Figure 5.9 is a Competition Analysis Guide that lists the types of questions you should answer when analyzing companies thought to be competitors. Comprehensive and thorough responses to these questions are necessary to have a reliable basis for making critical decisions about strategy. Finding answers to questions such as these will require the strategic planning team to enlist the assistance of a broad base of company personnel. Some of the information will be available electronically and through such documents as corporate annual reports. However, some of it will have to be collected the hard way—through old-fashioned research. For this information, take full advantage of your organization's personnel. Marketing, sales, human resources, adminis-

---

Competition Analysis Guide

- Directory information (Company name, address, CEO):
- Competing products/services and best features:
- Competitive Strategy:
- Common market segments:
  - Sales volume (annual)(past five years):
  - Percent of market:
  - Profits:
  - Annual growth (past five years):
- Organizational/product strengths:
- Organizational/product weaknesses:
- Pertinent developments, issues, or factors:

FIGURE 5.9 The adage, *know your enemy,* applies when assessing the competition.

trative, and technical personnel often know much about the competition from their individual perspectives. Some of your personnel might have worked for the competition at one time. Others might attend the same conferences or be members of the same professional organizations. Make sure that no one inside the organization who might have valuable insights is overlooked.

The information collected for each question is important in its own way. Consequently, it is important to develop answers that are thorough, comprehensive, and accurate. The better the information, the better the analysis. What is most important in answering each respective question is summarized in the following paragraphs.

Directory Information (Company Name, Address, CEO)

Where is the company located and who is in charge? The company's location might reveal potential issues relating to geographic advantages or disadvantages in certain market segments. The leadership style and level of commitment of the CEO can also be important. Has the CEO made a career of bouncing from company to company, or does he or she have a long-term commitment to this competitor? The more you know about the CEO's background, level of commitment, leadership style, and track record, the more you will know about how determined a competitor he or she will be.

Competing Products and Services and Their Best Features

It is important to know which of the competitor's products actually compete with yours in the market segments your organization wants to dominate.

Each product should be listed along with the features it carries that appeal most to customers. This information will help your organization decide if it should add features to its products, enhance existing features, or seek ways to minimize or simply avoid the competitor's best features.

Competitive Strategy

What is the competitor's overall competitive strategy? Is the competitor attempting to win on the basis of pricing (low-cost strategy)? Is the competitor attempting to differentiate its products and services? If so, on what basis? Is the competitor trying to reach a middle market through both pricing and differentiation? This information will help your organization in adopting its own competitive strategy.

Common Market Segments

Which market segments as refined in the previous step does this competitor have in common with your organization? This is where the most intense competition will occur. For every common market segment you need to know the competitor's sales volume, percent of market share, profits, and annual growth.

Organizational and Product Strengths

What are the strengths of the competitor's organization? What are the strengths of the competitor's products? Can your organization outperform the competitor in its areas of strength, or will you have to adopt strategies that avoid these strengths? Can your organization overcome the key strengths of the competitor's products, or will you have to adopt strategies that minimize these strengths while accentuating other aspects of your product? These are critical strategy-related questions.

Organizational and Product Weaknesses

What are the competitor's weaknesses? Are there ways your organization can exploit these weaknesses to gain a competitive advantage in the market segment in question? Does the competing product have weaknesses that might be exploited? The answers to these questions will become important when adopting strategies later in the process.

Other Pertinent Developments, Issues, or Factors

Are there developments, issues, or factors relating to this competitor that might help or hinder your organization's chances of winning in head-to-head competition? For example, is the competitor's CEO leaving under a cloud? Is the competitor being sued on a product liability issue? Has the competitor recruited a dynamic new marketing director or research scientist? Is the competitor's patent on an important process or product attribute about to expire? Any development, issue, or factor that might help or harm your organization from a competitive perspective is important and should be recorded here.

Summary

1. Part of the strategic planning process is about deciding what is possible. This decision cannot be made without a full and accurate assessment of the organization's financial condition. In assessing an organization's financial condition, at least three reports should be examined: statement of assets and liabilities, operating budgets, and the profit analysis statement.

2. Before conducting an assessment of the organization's strengths and weaknesses, the strategic planning team should take the time to prepare the organization. The CEO and senior executives should play leadership roles in carrying out this task. An effective approach is to conduct a series of small group meetings in which the point is made that an accurate assessment of organizational strengths and weaknesses is essential to future success. No negative use will be made of the information collected. It will be used to help the organization improve its competitive standing in the marketplace. Three effective methods for assessing organizational strengths and weaknesses are the functional unit method, value tree method, and key performance measures method.

3. The results of the assessment of organizational strengths and weaknesses are used in several ways: (1) when selecting the organization's overall competitive strategy; (2) when deciding how best to attack the competition; (3) when deciding where to avoid the competition; and (4) when deciding how to most effectively apply resources to strengthen weak areas and improve the organization's ability to compete.

4. A core competency that produces value is a mix of human talents and attitudes combined with organizational processes and technologies in ways that, when taken together, allow an organization to provide superior value to customers.

5. An organization should never assume it knows what its customers value most. Rather, the prudent approach is to look, listen, and ask. Methods for identifying what customers value most include the following: reviewing warranty records, reviewing customer service records, interviewing frontline personnel who have the most contact with customers, conducting customer interviews, conducting customer focus groups, soliciting written customer feedback, and recording, compiling, and reviewing customer complaints.

6. In deciding which organizational competencies are core value-producing competencies or if the organization even has any of these competencies, remember that all of the following characteristics must apply: (1) is critical to providing what customers value most; (2) combines human talent

and attitudes with organizational processes and technologies in a highly effective way; (3) effectively differentiates your organization's products and services from those of the competition; and (4) is sufficiently unique to be difficult for the competition to match, develop, acquire, or replicate.

7. The strategic planning process should aim for thriving in a competitive business environment, not just surviving. To thrive in a highly competitive environment, it may be necessary to refine the organization's markets to the point that it is the best in the business at providing value to customers in that tightly refined market.

8. The ideal market in a competitive environment is one in which the customers all value the same attributes of your product. To create such a situation, market segmentation is typically necessary. Markets can be segmented in a variety of ways. Some of the more commonly used methods are geographic, customer demographics, purchase method, and product type.

9. Once markets have been segmented, they should be analyzed carefully before being adopted. Questions to ask in analyzing markets are as follows: How is the market segment defined? What do customers in this market segment value most? What is the size of the market segment? What is our current share of this market segment? Who are our competitors in this market segment, and what is their market value? Are there trends, issues, or developments relating to this market segment we should consider?

Key Terms and Concepts

Assets and liabilities

Core competencies that produce value

Customer demographics segmentation

Functional unit method

Geographic segmentation

Key performance measures method

Market segmentation

Operating budgets

Product type segmentation

Profit analysis

Purchase method segmentation

S.W.O.T analysis

Strengths-and-weaknesses assessment

Surviving versus thriving

Threat factor

Value tree method

Review Questions

1. Explain how to assess the financial condition of an organization.
2. Summarize how to assess the strengths and weaknesses of an organization.

3. What are core competencies that produce value?

4. How do you distinguish an organizational strength from a core value-producing competency?

5. Explain how to conduct a market segmentation analysis for an organization.

6. What is the rationale for tightly refining the market segments in which an organization will compete?

7. What information do you need to thoroughly analyze a given market segment?

8. Explain how to analyze an organization's competition.

SIMULATION CASES FOR DISCUSSION

The following simulation cases are provided to generate additional thought and discussion about the principles explained in this chapter. Readers are encouraged to consider how the situations presented in these cases might apply to them and to discuss the cases with others interested in strategic planning and execution.

CASE 5.1   Are You Sure You Want the REAL Information?

Dale Cooper, CEO of Cooper Engineering Company (CEC), runs the company founded by his grandfather. But the market conditions he faces every day would have been inconceivable to his grandfather. Even Cooper's father did not have to contend with competition that is global in scope. But Dale Cooper does—every day. That's why he is taking his company through a comprehensive strategic planning process. The strategic planning team is in place and making good progress. Cooper is leading the team.

Much has already been accomplished. The strategic planning team is now to the point in the process where it must assess the company's financial condition. Cooper asked his comptroller to pull together the necessary financial information for the strategic planning team a week ago. Today, the comptroller gave Cooper a "sanitized" and upbeat summary of the company's financial condition.

"I don't want a marketing piece," said Cooper. "I want the real facts."

"Are you sure?" asked the comptroller. "Do you really want all the people on the strategic planning team to see numbers that nobody but you and I ever see?"

Discussion Questions

1. Do you see a problem with giving all members of the strategic planning team the real picture of the company's financial condition?

2. If you were Dale Cooper, what would you say to the comptroller in this case?

CASE 5.2 That's Not a Core Value-Producing Competency

Dale Cooper is having a debate with another member of the CEC strategic planning team. The vice president for engineering thinks CEC's new computer-aided design and drafting (CADD) system is a core value-producing competency. Cooper isn't sure he agrees.

"After all, accurate, attractive drawings are one of our products that customers really value," said the engineering vice president.

"I agree," said Cooper. "And the new CADD system certainly allows us to produce outstanding drawing packages. The system has given us an edge for right now, but how long will it be before the competition buys the same system and can operate it as well as we can?"

Discussion Questions

1. Do you agree with the engineering vice president in this case or Cooper?
2. How should the strategic planning team at CEC determine if its new CADD system represents a core value-producing competency?

CASE 5.3 I Think We Are Slicing Our Market Segments Too Thin

The vice president for marketing at CEC is concerned that the strategic planning team is refining the company's markets so tightly that CEC might lose customers. Dale Cooper argues that CEC needs to find out what it can do better than any other company and put all of its resources and efforts into developing those competencies. The strategic planning team members are split on this issue.

Discussion Questions

1. Join this debate. Take one side and defend your decision.
2. If the company's markets are sliced thin, how can the company make up for any customers that might be excluded?

Endnote

1. As quoted in Boone, L. E., *Quotable Business*, 2nd ed. (Random House, Inc., New York: 1999), 169.

Make Informed Predictions

The art of prophecy is very difficult, especially with respect to the future.[1]

—Mark Twain

OBJECTIVES

- Define the concept of informed predictions.
- Explain why the process of making informed predictions is so important to the strategic planning team.
- Explain how to make informed predictions as part of the strategic planning process.
- Describe how an organization can predict pertinent market behaviors.
- Explain how to identify potential opportunities and threats for an organization.

107

Strategic planning is about the future. Planning is by its nature a futuristic enterprise. It is a proactive process organizations use in an attempt to not just predict and respond to the future, but to shape it to their benefit. To shape the future, it is necessary to make predictions about what the future can hold if the organization adopts the right strategies and effectively carries them out. Remember, the best definition for strategic planning is *planning for the possible.*

CONCEPT OF INFORMED PREDICTIONS DEFINED

Do you ever watch the Weather Channel? If so, you are accustomed to the concept of forecasting. It's what meteorologists do. A forecast is nothing more than an informed prediction. An **informed prediction** is a supposition about future developments that is shaped by the research, experience, judgment, instincts, professional knowledge, and common sense of the individuals making it.

No human being can simply look into the future and know what is going to happen. However, we can look into the future and forecast what might happen or, better yet from the perspective of strategic planning, what could happen. When we do this, the more effectively we apply our experience, judgment, instincts, professional knowledge, and common sense to our research findings, the more informed and, hence, more accurate and credible our predictions are likely to be.

Return to the analogy of weather forecasting. When meteorologists prepare the forecasts they present on the Weather Channel, they are really making predictions about what future weather conditions will be. But their predictions are not just wild guesses. Rather, they are suppositions that have been subjected to careful scrutiny and shaped by research, experience, judgment, instincts, professional knowledge, and common sense.

The experience of weather professionals tells them that, when a given set of conditions exists in the environment, a certain type of weather will usually result. Their judgment and instincts give them an innate feel for what types of weather can be expected under what types of conditions. Thorough research reveals what has happened in the past under the same

Strategic Planning Tip

Key Point

An *informed prediction* is a supposition about future developments that is shaped by the research, experience, judgment, instincts, professional knowledge, and common sense of the individuals making it.

or similar conditions. Professional knowledge gives meteorologists a scientific basis for applying cause-and-effect paradigms. Finally, common sense overlays all aspects of the process as weather professionals attempt to tell us what conditions to expect in the days, weeks, and months ahead.

This concept of making informed predictions about what the future can hold for an organization is central to the strategic planning process. Members of the strategic planning team and others involved in the process must look into the future, conduct research, and then apply their experience, judgment, instincts, professional knowledge, and common sense in predicting what the future can hold for their organization provided the appropriate steps are taken at the appropriate time and in the appropriate ways. The process involves making informed predictions about market developments, future opportunities, and future threats.

Strategic Planning Tip

Key Point

For businesses operating in a competitive environment, making informed predictions about the future is a critical part of the strategic planning process. In fact, organizations that fail to make informed predictions about the future are setting themselves up for failure. The reason for this is simple. The world is always changing. It never stays the same. In business, the pace of change is even faster than in the world at large.

WHY MAKING INFORMED PREDICTIONS IS SO IMPORTANT

For businesses operating in a competitive environment, making informed predictions about the future is a critical part of the strategic planning process. In fact, organizations that fail to make informed predictions about the future are setting themselves up for failure. The reason for this is simple. The world is always changing. It never stays the same. In business, the pace of change is even faster than in the world at large.

The business environment changes, the marketplace changes, technology changes, the competition changes, and the customer changes. All of these changes affect an organization's ability to compete. Strategic planning is, by definition, a future-oriented enterprise. Without looking into the future and making informed predictions, an organization cannot adopt strategies to shape the future to its benefit. Shaping the future involves: (1) understanding what it is likely to hold for an organization; (2) knowing what to avoid as well as what to exploit; (3) knowing where to focus limited resources for the greatest possible benefit; (4) knowing what new products or product features to introduce as well as where and when;

STRATEGIC PLANNING PROFILE   Informed Predictions at John Deere

John Deere began as a one-person blacksmith shop in 1837 and was incorporated as Deere & Company in 1868. Since that time, John Deere has grown into an international business employing more than 40,000 people and conducting business in 160 countries. As a global competitor, making informed decisions about opportunities, threats, markets, and competitors is critical to the ongoing success of John Deere. Based on the best predictions the company's leadership team can make, John Deere has chosen to compete on a global basis in the following markets:

- Agriculture equipment for large and small farming operations
- Lawn and grounds care equipment for homeowners, groundskeepers, and golf and turf professionals
- Construction and forestry equipment
- Engines and power equipment

Source: John Deere Web site (http://www.deere.com)

(5) knowing what types of acquisitions might be appropriate and when to make them; and (6) knowing when to expand, pull back, invest, or divest. None of these things can be known without first making informed predictions about the future.

HOW TO MAKE INFORMED PREDICTIONS

You are already accustomed to making informed predictions. Before taking a long business trip you make informed predictions about such things as the types and amounts of clothing you will need, food, lodging, and transportation. With regard to clothing for example, you might ask yourself questions such as these: Will I need three suits or four? Will I need cold-weather clothing (e.g., gloves, overcoat, hat, snow boots)? Will I need leisure wear? Will I need athletic clothes for workouts?

To answer these questions, you must make certain predictions. As to how many suits to bring, you have to predict how many meetings you will have to attend and what the dress code for those meetings will be. As to cold-weather gear, you have to conduct enough research to predict what the weather conditions will be at your destination. As to leisure wear, you must predict any events or activities outside of your meeting that you might need or just want to attend. As to athletic clothing, you have to predict the amount of unscheduled time you will have available and whether or not there will be fitness facilities available. If you have experience traveling to business meetings, you probably have developed the instincts, judgment, common sense, and other characteristics needed to make informed predictions about the clothing issue.

Making informed predictions during the strategic planning process is similar to planning for the business trip in this scenario. The first step is to identify the variables to be considered. The second step is to conduct any research that needs to be done. The third step is to apply your experience, judgment, instincts, professional knowledge, and common sense to interpreting the research findings. In this way, you make informed predictions.

Avoid Common Planning Pitfalls

One of the first steps in predicting the future is to identify trends from the past. This is one of the essential research tasks that must be completed as part of the prediction process. Historic trends can be very helpful in establishing a foundation upon which to build predictions. However, one of the most common pitfalls when making predictions is to assume that because markets, customers, or the business environment behaved in certain ways in the past, they will continue the same behavior patterns in the future. They might, but they might not. For this reason, when examining historic behavior trends, don't fall into the trap of making the misguided assumption of automatically accepting historic trends as accurate indicators of the future. Rather, consider other cultural, technological, socioeconomic and sociopolitical factors and how they might affect the trend lines.

For example, hypothesize that your company manufactures communications equipment for use in commercial airliners. For the sake of discussion, let's say that the pertinent historic trend lines point to continued growth in the manufacture of new commercial jets. You might interpret the trend lines as telling you to go ahead with a major corporate expansion to meet the growing need for your product. However, before making such a huge investment, the strategic planning team would want to consider the potential impact of international terrorism on future air travel. Will the trend lines that have

---

Strategic Planning Tip

Key Point

One of the first steps in predicting the future is to identify trends from the past. This is one of the essential research tasks that must be completed as part of the prediction process. Historic trends can be helpful in establishing a foundation upon which to build predictions.

historically shown steady growth continue to show growth once the threat of terrorism has been factored in? Are there other cultural, technological, socioeconomic, or sociopolitical factors that should be considered?

Another common pitfall is tunnel vision. When predicting the future, it is best to hypothesize as many potential scenarios as possible. Each scenario is then discussed, debated, subjected to analysis, and rated in terms of its viability. However, even the most intelligent people can become so focused on the status quo or so invested in how things are at the moment that they develop tunnel vision and simply refuse to give serious consideration to other scenarios. This is a dangerous pitfall. Consider the following examples:

1. Thomas J. Watson, founder and first president of IBM Corporation, once said: "There is a world market for about five computers."[2]
2. Octave Chanute, French aviation pioneer, once said: "The flying machine will eventually be fast; they will be used in sport but they should not be thought of as commercial carriers."[3]
3. American inventor Lee DeForest once said: "While theoretically and technically television may be feasible, commercially and financially I consider it an impossibility, a development of which we need waste little time dreaming."[4]

Each individual quoted here was a bright, successful person. Each eventually made a significant impact on our world. But at the time in their lives when they said the words quoted here, each suffered from an acute case of tunnel vision. These individuals made the mistake of basing their predictions on the existing conditions of the moment. They failed to consider cultural, technological, socioeconomic, and sociopolitical factors that might change conditions in the future.

Most people have at least a tendency toward tunnel vision. For this reason, it is important to lay out as many potential scenarios of the future as possible during this phase of the strategic planning process and to think as creatively as possible when doing so. Once the various possible scenarios have been

---

Strategic Planning Tip

Key Point

When predicting the future, it is best to hypothesize as many potential scenarios as possible. Each scenario is then discussed, debated, subjected to analysis, and rated in terms of its viability. However, even the most intelligent people can become so focused on the status quo or so invested in how things are at the moment that they develop tunnel vision and simply refuse to give serious consideration to other scenarios.

painted, they can be evaluated for viability. This is where such characteristics as experience, judgment, instinct, and common sense come into play.

PREDICTING PERTINENT MARKET BEHAVIORS

Business markets are like the weather—they are always changing. It is important to understand this simple truth because market changes can affect your organization's ability to compete. Observe your markets in the same way that scientists observe the environment. Scientists are always studying environmental conditions in an attempt to predict hurricanes, floods, earthquakes, tornadoes, volcanic eruptions, and other potentially catastrophic events.

Scientists know that on some level, everything in the environment is connected. A small change in the environment today might begin a series of events that will lead to major changes in the future. For example, if the rain forest is depleted in South America, the weather might be affected in North America. Smog released into the atmosphere in Los Angeles might eventually affect the ozone layer over Antarctica, which in turn, might affect the climate of Argentina. Looking not just for environmental developments but also for connections between and among these developments is important to scientists who make predictions about the environment. In other words, scientists look for cause-and-effect relationships in the environment.

Businesses should follow the lead of scientists and apply their model to studying the business environment. Technical professionals involved in strategic planning must observe their business environment with an eye to cause-and-effect relationships. A trucker's strike in Chicago might affect the ability of a business in Atlanta to secure the supplies it needs to operate. A war in the Middle East might cause the price of gasoline to skyrocket. When this happens (the cause component), few businesses will be spared as the economic ramifications (the effect component) begin to cascade down through the economy. If the Occupational Safety and Health Administration (OSHA) adopts a new standard in Washington, D.C., the cost of doing business might increase for companies throughout the country.

Strategic Planning Tip

Key Point

Business markets are like the weather—they are always changing. It is important to understand this simple truth because market changes can affect your organization's ability to compete. Observe your markets in the same way that scientists observe the environment.

Strategic Planning Tip

Key Point

The business environment is just like the natural environment in that a seemingly unrelated event that occurs hundreds of miles away can initiate a series of resultant events that, over time, have a profound effect on the ability of your organization to compete. Consequently, it is important for organizations to stay in close touch with the market segments that make up their business environment and to continually analyze the markets in the same way that scientists analyze the natural environment.

The business environment is just like the natural environment in that a seemingly unrelated event that occurs hundreds of miles away can initiate a series of resultant events that, over time, have a profound effect on the ability of your organization to compete. Consequently, it is important for organizations to stay in close touch with the market segments that make up their business environment and to continually analyze the markets in the same way that scientists analyze the natural environment.

Scientists divide the natural environment into segments as a way to simplify the processes of monitoring and predicting. For example, some scientists focus on the oceans, lakes, and rivers of the world; others on the air; and still others on the earth and what is under it. Technical professionals involved in strategic planning should do the same thing. In Chapter 5, the concept of breaking a market into tightly defined segments was explained. This chapter explains how to examine each of these segments from the perspective of future developments.

The strategic planning team must answer three questions for each of the organization's market segments:

1. What pertinent business developments are likely to occur that might affect—positively or negatively—the organization's ability to compete in this market segment?
2. What cultural, socioeconomic, or sociopolitical developments are on the horizon that might affect this market segment?
3. What business, cultural, socioeconomic, or sociopolitical developments could negate this segment as a viable market for the organization or even eliminate it?

Pertinent Business Developments

Environmental scientists who monitor weather conditions in an attempt to predict hurricanes have specific factors they focus on such as barometric pressure, water temperature, and upper-level wind patterns. These are critical factors in the formation of hurricanes. Technical professionals involved in strategic planning should follow the lead of these environmental scientists when trying to predict pertinent business developments relating to a given market sector. They too should focus on specific critical factors. Factors to consider when predicting pertinent business developments are customer desires and profitability potential.

Customer Desires

The functional question relating to customer desires is this: What will customers want from us and our product in the future? Although it might seem that you would need a crystal ball to answer this question, you don't. The future desires of customers can be predicted by considering trends from the past. Customers have always desired products that provide ever-increasing value. That value might be defined as more convenience; enhanced comfort; greater adaptability; ease of installation, integration, and maintenance; more reliability; smaller size; better transportability; enhanced safety; and any other factor that will make the product easier to purchase, use, maintain, and replace.

When considering the issue of customer demands, remember that the issue is value. Anything that adds value to your product for customers will be a welcome addition. Just consider developments to date in such markets as automobiles, telecommunications, and personal computers. Customers have continually demanded more convenience, reliability, ease of operation, system integration, enhanced safety, and convenience. Customer demands on your product will follow similar patterns. Just put yourself in your customers' shoes and ask the following question: What would make this product more valuable to me?

Strategic Planning Tip

Key Point

When considering customer demands, remember that the issue is value. Anything that adds value to your product for customers will be a welcome addition.

Profitability Potential

Having considered the future desires of customers, the strategic planning team must then ask itself: Can our organization provide superior value to customers in the future and make an acceptable profit? To answer this question, it will be necessary to once again make internal and external observations, but this time with an eye to the future rather than the present.

The functional questions are these: (1) Is the number of competitors in this market segment likely to increase, plateau, or decrease in the future? (2) If the number of competitors increases, can the organization provide superior value in satisfying customer desires? (3) Can the organization outperform the competition, provide superior value, and still maintain an acceptable profit margin?

Market Development

Once the strategic planning team has considered what customers will demand of the organization and its products in the future, it should consider the next obvious question: How many customers will make these demands? The issue is market development. Will the market segment in question grow or shrink in the future? This question pertains to the market segment for all competitors interested in it, not just for your company. Once you have refined a market into a tightly defined segment, your organization's percent of market share will depend on its performance in providing superior value to customers.

The senior executives who serve on the strategic planning team might know or might think they know what the key drivers are for the market segment in question. However, making assumptions without first doing your homework can be a serious mistake for those engaged in strategic planning. Rather than assume, it is wise to investigate. Study past socioeconomic, sociopolitical, business, and cultural events and trends to determine what effect they had on the organization's markets. Look for cause-and-effect relationships. Making these connections can be like peeling an onion. Every time you peel away one layer, there is always another, and each successive layer can bring tears. However, the benefits of undertaking this task will eventually be reaped in the form of a better strategic plan for the organization.

Strategic Planning Tip

Key Point

Study past socioeconomic, sociopolitical, business, and cultural events and trends to determine what effect they had on the organization's markets. Look for cause-and-effect relationships. Making these connections can be like peeling an onion. Every time you peel away one layer, there is always another, and each successive layer can bring tears. However, the benefits of undertaking this task will eventually be reaped in the form of a better strategic plan for the organization.

Begin with the most basic economic issues such as inflation, recession, fluctuations in oil prices and other obvious factors, and look for cause-and-effect relationships. In the past, when this event occurred (cause), our market responded as follows (effect). Then, when you have solid trend lines relating to cause-and-effect relationships for a given factor, such as inflation, study what credible economists say about the likelihood of that factor occurring again. Repeat the process for all pertinent economic factors.

After considering economic issues, move on to demographics. What demographic issues might affect the market segment? For example, if the product is age sensitive, what do the demographic trends for the age groups in question show—growth or shrinkage? If the market segment has been defined geographically, what are the demographic trends for the geographical area in question?

Once the applicable demographic issues have been considered, peel another layer off the onion and consider such things as natural disasters, terrorist attacks, and political turmoil. Is the market sector in question sensitive to hurricanes, earthquakes, tornadoes, floods, or terrorist-induced disruptions to travel, power supplies, water systems, or other necessities of life? Will political turmoil such as mini-conflicts around the world cause the market segment to grow or shrink? What is the likelihood of any of these events occurring in the future? Of course, predicting the occurrence of a natural disaster or a war is a difficult undertaking at best. However, some areas of the world are more prone to certain natural disasters than others. For example, an organization that makes Florida, California, or Oklahoma part of a geographically defined market segment has no choice but to factor the possibility of hurricanes, earthquakes, and tornadoes into its strategic planning process. An organization that does business with countries in the Middle East must factor the potential for war into its plans.

Market drivers are many and varied depending on the nature of the organization in question and its products. Consequently, no attempt is made here to present a comprehensive list of factors that organizations should consider when predicting market development. Strategic planning teams for individual organizations must determine what factors are pertinent for those organizations. However, in deciding which factors to consider— beyond the obvious factors mentioned in this section—it is important to: (1) think broadly and creatively, (2) identify how any factors considered have affected the market in the past, and (3) keep peeling the onion while looking for cause-and-effect relationships.

Market Viability

The future viability of a given market segment might be the most difficult concept the strategic planning team has to predict. During the American Civil War, President Abraham Lincoln was persistently pestered by a man

---

Strategic Planning Tip

Key Point

Strategic planning teams for individual organizations must determine what factors are pertinent for those organizations. However, in deciding which factors to consider it is important to: (1) think broadly and creatively, (2) identify how any factors considered have affected the market in the past, and (3) keep peeling the onion while looking for cause-and-effect relationships.

---

who manufactured uniforms for the Union Army. This man wanted advance warning of the war's conclusion so that he wouldn't be left holding warehouses full of uniforms for which there was no longer a market. He got no response from President Lincoln, but one can certainly understand why he tried. For any market segment, there might be some factor quietly lurking just beneath the surface that could rob that sector of its viability or eliminate it altogether.

One of the best-known examples of an industry failing to consider future market viability is the Swiss watch-making industry. The inventor of the quartz crystal-powered watch took his invention to several Swiss watchmakers only to be turned away. They weren't interested because they could not conceive of customers preferring this new and different technology to their traditional wind-up technology. The Swiss watchmakers thought that because people had always bought wind-up watches, they would continue to do so. The rejected inventor then took his new technology to the Japanese, and the rest is history.

The Swiss watchmakers who rejected the inventor of quartz-powered watches failed to understand that they were looking at the watch-making technology of the future, a technology that would transform the watch market in ways that would undermine the viability of markets that were hundreds of years old. As a result, Swiss watchmakers saw the market for wind-up watches quickly transformed from a huge, broad-based market to a much smaller specialty market.

Technical professionals serving on strategic planning teams need to be especially sensitive to developments that could threaten or even eliminate their markets. For example, some manufacturers of large gas-guzzling automobiles in the 1970s failed to predict the impact of skyrocketing oil and gasoline prices and, as a result, saw Japanese manufacturers of smaller, fuel-efficient cars make significant inroads into markets the gas guzzlers had traditionally dominated.

Motorola was almost run over in the marketplace when Nokia was quicker to recognize the potential for digital cellular telephone technology. At the time, Motorola dominated the cellular telephone market by produc-

Strategic Planning Tip

Key Point

Technical professionals serving on strategic planning teams need to be especially sensitive to developments that could threaten or even eliminate their markets. For example, some manufacturers of large gas-guzzling automobiles in the 1970s failed to predict the impact of skyrocketing oil and gasoline prices and, as a result, saw Japanese manufacturers of smaller, fuel-efficient cars make significant inroads into markets the gas guzzlers had traditionally dominated.

ing high-quality analog phones. Instead of asking themselves what effect the new digital technology might have on the viability of their cellular telephone market, Motorola officials continued to make better and better analog phones. Eventually, this failure cost the company dearly.

At one point in its history, computer giant IBM was rocked to its foundations when key IBM officials failed to recognize that technological developments would convert the world of business from mainframe to personal computers. The list of companies that either failed or at least had to dig themselves out of deep holes because they overlooked the issue of future market viability is a long one that contains the names of some of the most revered companies in the history of business.

Factors that might detract from the viability of a given market segment are many, varied, and different for different types of organizations. Consequently, it is not possible to do more at this point than encourage strategic planning teams to study the issue of future market viability carefully and to consider every possible factor that could have an adverse effect on their market segment. One of the best ways to do this is to ask all members of the strategic planning team to read business cases about companies that failed

Strategic Planning Tip

Key Point

For the purpose of strategic planning, an opportunity should be viewed as any challenge that, if properly handled, will improve the organization's ability to compete and win on a consistent basis. A threat should be viewed as any challenge that, if ignored or improperly handled, will detract from the organization's ability to compete and win.

to consider future market viability and to apply what is learned from the cases to their planning process. Business literature is replete with such cases.

IDENTIFYING OPPORTUNITIES AND THREATS

Chapter 5 introduced the concept of the S.W.O.T. analysis. To review, S.W.O.T. stands for *strengths, weaknesses, opportunities,* and *threats.* The process for identifying organizational strengths and weaknesses was explained in Chapter 5. In this section, the process for identifying opportunities and threats is explained.

For the purpose of strategic planning, an opportunity should be viewed as any challenge that, if properly handled, will improve the organization's ability to compete and win on a consistent basis. A threat should be viewed as any challenge that, if ignored or improperly handled, will detract from the organization's ability to compete and win. You can probably sense from these definitions that opportunities and threats are often just opposite sides of the same coin. Often what determines whether a given challenge is an opportunity or a threat depends on how the organization handles the situation.

When attempting to identify opportunities and threats, the strategic planning team might need help. The author recommends the formation of several ad hoc groups, each one led by a member of the strategic planning team. Each ad hoc group brainstorms to identify as many potential opportunities and threats as possible. Creative thinking is encouraged. The ideas of each ad hoc group are summarized and submitted to the strategic planning team, which in turn develops a comprehensive summary of the input from all groups. The strategic planning team then rates each opportunity and threat contained in the master summary according to its potential impact on the organization.

The Brainstorming Process

During the brainstorming sessions, members of the strategic planning team serve as facilitators who keep the ideas flowing by drawing out participants. Members of the ad hoc groups are encouraged to share any idea that comes to mind with no limits and no judgments as to viability from other participants. At this point in the process, all ideas are considered valid. Participants are not allowed to make judgmental comments or to evaluate proposed ideas.

One member of the group is asked to serve as the recorder. This individual records all ideas on a flip chart or marker board as they are proposed. It's important that all participants be able to see the ideas when they are recorded. This serves two purposes. First, it keeps participants from repeating the same ideas. Second, one idea from one participant might trigger other ideas from others.

Brainstorming, if handled well, can be an excellent vehicle for collecting input and ideas from a broad base of personnel. However, potential problems must be effectively dealt with to gain the benefits of brainstorming. One problem is that people are sometimes reluctant to share their ideas in a group out of embarrassment or shyness. Consequently, it is important for the facilitator to draw participants out and to ensure that judgmental comments, facial expressions, and body language don't inhibit the flow of ideas.

Another potential problem is the tendency of some people to just go along with the group. These individuals typically have ideas of their own but rather than share them, they hold back and go along with what seems to be the group's consensus. People who hold back in such cases are more concerned with agreement than with ideas. The facilitator must be sensitive to this phenomenon and work to overcome it. One strategy that works well is to distribute 3×5 cards and ask all participants to write an idea on their card. The cards are then collected by the facilitator and the ideas are written on the flip chart or marker board by the recorder. Usually, by the time the group has gone through this written process a couple of times, participants will be ready to open up.

Factors to Consider When Identifying Opportunities and Threats

The quality of the input provided in the brainstorming sessions can be enhanced by giving some structure to the process. Figure 6.1 is a guide that can be used to organize the thinking of individuals who participate in brainstorming sessions for identifying opportunities and threats. Facilitators should make it clear to participants that Figure 6.1 is just a guide to trigger their thinking and organize their input. Ideas that do not fall into one of the categories in this guide should be both accepted and encouraged.

Internal Processes

Typically both opportunities and threats are associated with an organization's internal processes. In a competitive business environment an organization cannot dominate a market segment consistently unless it improves its key processes continually. This includes all of the various processes set forth early in this book in the value tree. If, for example, the organization's conversion processes are not efficient and effective enough to outperform those of the competition, they must be improved. This challenge, as is so often the case, can be viewed as either an opportunity or a threat. Regardless of the strategic planning team's perspective in this regard, the organization's conversion processes must be improved.

Competition

Are current competitors failing or struggling? If so, here is an opportunity to add market share. Are new competitors entering the organization's key market segments? If so, here is a threat that must be dealt with by outperforming

---

Factors to Consider When Identifying
OPPORTUNITIES AND THREATS

1. **Internal Processes**
   - Input processes
   - Conversion processes
   - Ouput processes
   - Sales/marketing processes
   - Support processes
2. **Competition**
   - Current competitors
   - New competitors
3. **Products/Services**
   - New products/services
   - Present products/services
4. **Markets**
   - New markets
   - Present markets
5. **Aquisitions/Mergers**
   - Purchase a competitor
   - Purchase a new capability
   - Be purchased
6. **Cultural/Socio-Economic/Sociopolitical Issues**

FIGURE 6.1   Guide for identifying opportunities and threats.

the newcomer and providing such superior value to customers that the new competitor is unable to gain a foothold in the market.

Products and Services

Are there problems with existing products? If so, here is an opportunity to improve the organization's competitive standing or a threat that could undermine it. Is the market segment the organization has targeted ready for a new product? If so, here is an opportunity to be the first on the market with a product that meets this need. Here also is a threat that if the organization doesn't provide the new product, a competitor will.

Markets

Do new markets have potential for the organization? If so, here is an opportunity to expand the organization's customer base. Are changes occurring in existing market segments? If so, here is an opportunity to respond faster and better than can your competitors. Here also is a threat that if the organization fails to respond faster and more effectively, the competition will.

Acquisitions and Mergers

Is there an opportunity to expand the customer base by purchasing a competitor? If so, here is an opportunity to grow while simultaneously removing a competitor. Here also is a threat that if the organization chooses not to purchase the company in question, another competitor might. Is there an opportunity to purchase a company that would immediately give the organization a new capability?

For example, a company might find itself attracting more and more customers who need the organization to have a strong machining capability. The organization has a few machines it has always used to do odd jobs, but machining has not been a major part of its conversion processes. It now needs to be, and there is an excellent machining company nearby that is up for sale. This situation represents an excellent acquisition opportunity for the organization.

Has the company attracted the attention of several larger organizations that are considering a hostile takeover? If so, here is a threat that must be addressed.

Cultural, Socioeconomic, and Sociopolitical Issues

This category of opportunities and threats requires the most creative thought. In the various steps that have led up to this point, the strategic planning team has looked at cultural, socioeconomic, and sociopolitical issues several times. Consequently, the team's thinking should already be attuned to the task at hand.

Strategic Planning Tip

Key Point

If the organization's product is culturally sensitive—that is, its perceived value by customers is subject to cultural trends—considering cultural threats and opportunities is imperative. But even if the organization's product does not appear to be culturally sensitive, don't overlook this area. Sometimes secondary and tertiary cause-and-effect relationships can affect the organization's competitive position.

If the organization's product is culturally sensitive—that is, its perceived value by customers is subject to cultural trends—considering cultural threats and opportunities is imperative. But even if the organization's product does not appear to be culturally sensitive, don't overlook this area. Sometimes secondary and tertiary cause-and-effect relationships can affect the organization's competitive position. For example, American culture constantly demands such characteristics as smaller, faster, more convenient, greater comfort, easier accessibility, and other factors that make life easier. Can the cultural demands for these characteristics affect attitudes toward the product?

Socioeconomic trends are the easiest to predict in this step of the process because so much help is available. Predicting future socioeconomic trends is what economists do, and their findings are widely published and openly discussed in a variety of forums. The key here is to view the predictions of economists from the perspective of how economic trends might affect your organization and its products.

Predicting sociopolitical issues requires members of the strategic planning team to get well informed about issues outside of the field of business. Political unrest in the Middle East, fluctuating oil prices, new monetary developments in the European Union, opportunities created by the fall of communism in the former Soviet Union, the ongoing struggle with North Korea over nuclear weapons, the political tendency toward a proliferation of government regulations, and increasing levels of oversight from government agencies are just some of the many types of sociopolitical issues the strategic planning team should consider.

Sociopolitical Issues: A Case in Point

A good example of why it is so important to study sociopolitical issues carefully and consider how they might affect the organization's competitive position is the on-again–off-again debate in Congress over the ergonomics standard proposed by the Occupational Safety and Health Administration (OSHA). This standard, if it is ever passed, will carry a substantial price tag for midsized and large technology companies. Meeting the mandates of the standard would require a huge investment that might have little or no return for businesses. Consequently, business groups and professional organizations have vigorously opposed OSHA's proposed ergonomics standard. The issue has been a real political football.

During the presidency of Bill Clinton, Congress steadfastly refused to approve OSHA's recommended ergonomics standard. President Clinton wanted the standard passed. Consequently, just prior to leaving office, he exercised the privilege of the presidency and signed an executive order putting the various elements of the standard in place as a rule. Once George Bush was inaugurated as President, Congress used the Congressional Review Act to overturn Clinton's executive order. OSHA responded by asking businesses to voluntarily adopt the various elements of the standard. As of this writing, complying with the ergonomics standard is a voluntary undertaking. This is just one sociopolitical issue businesses should be monitoring. There are many others.

City councils, county commissions, state legislatures, and Congress have the authority to pass laws, statutes, rules, and regulations that can knowingly or unknowingly have a detrimental effect on the ability of organizations to do business. Consequently, it is important to stay in touch with the issues placed before these governmental bodies. Perhaps the best way to stay in touch with sociopolitical issues is through membership in various business and professional organizations. Business and professional organizations keep close tabs on Congress and let their members know when potentially negative or positive legislation is being considered.

Summarizing Opportunities and Threats

Once the strategic planning team has collected the organization's best thinking concerning opportunities and threats, the input should be summarized and each item on the final list should be rated according to its probability and potential impact. Figure 6.2 is an example of a summary of

ABC Company Summary
OPPORTUNITIES AND THREATS

Opportunities	Probability	Impact
1. Jones Company will file for Chapter 11 protection.	High	Highly Positive
2. Our new XYZ product scored high in market tests and may be ready to introduce 6 months early.	High	Highly Positive
3. Congress might pass tax incentives for product research.	Moderate	Positive

Threats	Threat Level	Priority
1. Two key conversion processes are outdated.	High	High
2. Internet marketing needs improvement.	High	High
3. Potential hostile takeover.	Low	Moderate

FIGURE 6.2   Opportunities and threats should be rated and ranked.

opportunities and threats for a hypothetical organization. This hypothetical company used the guide in Figure 6.1 to structure its deliberations.

Notice that each opportunity on the list is rated according to *probability* and *impact*. The probability rating (high, moderate, or low) describes the likelihood of this opportunity actually materializing. The impact rating (highly positive, positive, somewhat positive) describes the effect the opportunity could have on the organization's competitive position in the marketplace.

Each threat on the list is rated according to the *threat level* and its *priority*. The threat-level rating (high, moderate, low) describes the magnitude of the threat to the organization. The priority rating (high, moderate, low) describes how important dealing with the potential threat is to the organization. This determination is based on the potential harm of the threat if it is ignored and then actually materializes.

Three opportunities are listed in Figure 6.2. The first falls in the category of competition (from Figure 6.1). The second falls in the category of products/services. The third opportunity falls under the heading of sociopolitical issues. The strategic planning team determined that there is a high probability that the Jones Company, a competitor, will need to file for Chapter 11 protection from bankruptcy. If this happens, it will have a highly positive effect on the ABC Company, which will of course attempt to win over that share of the market historically controlled by the Jones Company.

The strategic planning team determined that there was a high probability that its new XYZ product—a product that had done well in market tests—would be ready to introduce six months earlier than originally planned. This will have a highly positive effect because it will allow ABC Company to put a new, potentially popular product on the market ahead of the competition, thereby gaining a foothold in the marketplace on which it can then build.

The strategic planning team has been monitoring a specific piece of legislation under consideration by Congress that would award companies tax incentives for product research in certain areas. The team determined that it was only moderately likely that the bill in question would pass, but that it will have a positive effect for ABC Company if it does.

Three threats are listed in Figure 6.2. The first threat falls in the category of internal processes (from Figure 6.1). The strategic planning team determined that two of the company's key conversion processes were in need of updating and that failure to complete the upgrade represented a high-level threat. Consequently, the strategic planning team made upgrading the conversion processes in question a high priority.

The strategic planning team also determined that ABC Company needed to do a better job of marketing its products over the Internet. Failing to do so was determined to represent a high-level threat. Consequently, enhancing the company's Internet marketing was assigned a high priority. Finally, there was some concern over the possibility of a hostile takeover of ABC Company. However, the strategic planning team determined that the likelihood of this happening was low. Even so, the team assigned monitoring this situation a moderate priority so that the company would not be taken by surprise in the future by a corporate raider.

Summary

1. An informed prediction is a supposition about future developments that is shaped by the research, experience, judgment, instincts, professional knowledge, and common sense of the individuals making it. Strategic planning requires the strategic planning team to make informed predictions about market developments, future opportunities, and future threats.

2. The rationale for making informed predictions about the future is simple. The world is always changing. It never stays the same. In business, the pace of change is even faster than in the world at large. To function in such an environment, businesses must attempt to predict and shape future developments.

3. Making informed predictions during the strategic planning process involves identifying the variables involved, conducting research into the variables, and applying experience, judgment, instincts, professional knowledge, and common sense in interpreting the research findings.

4. Common pitfalls of strategic planning teams attempting to make informed predictions include the following: assuming that because markets, customers, or the business environment behaved in certain ways in the past, they will continue the same behavior patterns in the future, and letting tunnel vision limit the possible scenarios considered.

5. Technical professionals should monitor the business environment with an eye to cause-and-effect relationships. An event that might seem unrelated can actually turn out to have a profound effect on an organization. Consequently, the strategic planning team should answer three key questions about each of its market segments: (1) What pertinent business developments are likely to occur that might affect—positively or negatively—the organization's ability to compete in this sector? (2) What cultural, socioeconomic, or sociopolitical developments are on the horizon that might affect this market segment? (3) What business, cultural, socioeconomic, or sociopolitical developments might negate this market segment as a viable market for the organization?

6. Pertinent business development factors to consider when predicting market behaviors in a given market segment include customer desires and profitability potential. The pertinent market development factor to consider is this: Will the market segment in question grow or shrink in the future? When considering future market viability, the strategic planning team

should look for developments that could undermine or even eliminate the market segment in question.

7. The brainstorming process is an effective way to identify opportunities and threats that could manifest themselves in the future. When attempting to identify potential opportunities and threats, it helps to consider such categories of concern as the following: internal processes, competition, products/services, markets, acquisitions/mergers, and cultural, socioeconomic, and sociopolitical issues.

Key Terms and Concepts

Acquisitions and mergers
Brainstorming process
Common planning pitfalls
Competition
Cultural, socioeconomic, and
  sociopolitical issues
Customer desires
Informed predictions
Internal processes

Market development
Market viability
Markets
Opportunities
Pertinent business developments
Products and services
Profitability potential
Threats

Review Questions

1. Define the concept of informed predictions.
2. Explain why making informed predictions is so important to the strategic planning process.
3. Describe the process for making informed predictions.
4. What are the common pitfalls to avoid when making informed predictions?
5. Explain the factors to consider when predicting pertinent business developments.
6. Explain the factors to consider when predicting market development in the future.
7. Explain the factors to consider when predicting future market viability.
8. Describe how to use brainstorming to identify opportunities and threats for an organization.
9. What main factors should be considered when identifying opportunities and threats?

SIMULATION CASES FOR DISCUSSION

The following simulation cases are provided to generate additional thought and discussion about the principles explained in this chapter. Readers are encouraged to consider how the situations presented in these cases might apply to them and to discuss the cases with others interested in strategic planning and execution.

CASE 6.1 Why Do This—We Can't Predict the Future

"Why do this—we can't predict the future?" This statement was made by the vice president for engineering at CAM Tech, Inc. (CTI). He was referring to the step in the strategic planning process in which the CTI strategic planning team must make informed predictions.

"Maybe we can't predict the future, but we can attempt to learn as much as possible about what might happen," said CTI's vice president for marketing.

"I'd like us to go beyond predicting just what might happen," said CTI's CEO. "I want us to consider what could happen, both good and bad, depending on our actions as a company."

Discussion Questions

1. Take the side of the either the vice president for engineering or the vice president for marketing and make your case.
2. Discuss what the CEO means by predicting what "could" happen.

CASE 6.2 The Market for Our Products Is as Predictable as the Weather

"The market for our products is as predictable as the weather," said CTI's vice president for engineering. "By that I mean we cannot possibly predict how it's going to behave in the future."

"I disagree," said the vice president for manufacturing. "If we know what to look for we can make informed predictions that will have an acceptable level of credibility, even if they are not completely accurate."

Discussion Questions

1. Can an organization predict the future behavior of its market segments with an acceptable degree of accuracy?
2. Discuss how an organization might go about making informed predictions about the future behavior of selected market segments.

CASE 6.3 Predicting Opportunities and Threats Makes Sense

"Finally, we are talking about something I can relate to," said CTI's vice president for engineering. "Predicting opportunities and threats makes sense."

"I'm glad to hear you say that," said the company's CEO. "What changed your mind?"

"Predicting opportunities and threats is different than the other predictions we have been trying to make. I hear things at professional meetings. I read the newspaper and follow current events in the news. I read my professional journals and talk to colleagues from other companies. I have something to go on."

Discussion Questions

1. Do you agree with CTI's vice president for engineering that predicting opportunities and threats is a less difficult undertaking than predicting the behavior of market segments? Explain your position.

2. Discuss ways that you can stay well enough informed to make credible predictions about opportunities and threats to an organization.

Endnotes

1. As quoted in Boone, L. E., *Quotable Business*, 2nd ed. (Random House, Inc., New York: 1999), 309.

2. Ibid., 310.

3. Ibid.

4. Ibid.

Adopt a Strategic Emphasis and Competitive Strategy

Business is like war in one respect. If its grand strategy is correct, any number of tactical errors can be made and yet the enterprise proves successful.[1]

—Robert E. Wood

OBJECTIVES

- Explain why it is important to adopt a strategic emphasis for the organization.
- Summarize the potential areas of strategic emphasis available to an organization.
- Demonstrate how to select the strategic emphasis that is best for the organization.
- Explain how to select the competitive strategy that is best for the organization.

Strategic planning is a process that helps organizations answer three specific questions that are critical for any company that operates in a competitive business arena:

1. Who are we?
2. Where are we going?
3. How will we get there?

All of the steps in the process completed thus far have been in preparation for what happens in this step—adopting a strategic emphasis and a competitive strategy. Think back over what has occurred up to this point in the process. The strategic planning team has:

1. Made its members knowledgeable of the concepts of competition and the need to have competitive advantages.
2. Made its members knowledgeable concerning the concept of value and how it affects the organization's ability to compete.
3. Assessed the financial condition of the organization.
4. Identified the organization's strengths and weaknesses.
5. Identified the organization's core competencies that produce value or what those competencies need to be in order to be competitive.
6. Begun looking at markets with a mind to refining them into tightly defined segments.
7. Begun analyzing the organization's competitors to identify their areas of strength and vulnerability.

All of the work accomplished, all of the information collected, and all of the analysis completed up to this point in the process has been to provide the background and context necessary to complete the current step and those that remain in the strategic planning process.

The strategic planning team will now use the background information collected along with new information to adopt a strategic emphasis that will answer the first two foundational questions of strategic planning:

Strategic Planning Tip

Key Point

An organization's strategic emphasis tells who it is and where it's going. These two things are important to know. Executives cannot effectively lead an organization without knowing who it is and where it's going.

(1) Who are we? (2) Where are we going? The strategic planning team will also answer the third question in the process: How will we get there? This question is answered by adopting a competitive strategy for each market segment to be pursued.

IMPORTANCE OF ADOPTING A STRATEGIC EMPHASIS

An organization's strategic emphasis tells who it is and where it's going. These two things are important to know. Executives cannot effectively lead an organization without knowing who it is and where it's going. Investors have no viable basis for deciding whether or not to invest in the organization unless they know who it is and where it's going. Employees cannot fully understand how best to help the organization unless they know who it is and where it's going.

The value of using the segmentation process to refine broad markets into tightly defined market sectors is explained in Chapter 6, and its importance is emphasized. The competitive benefits of being a champion in a tightly defined market sector, rather than just another player in a broadly defined market, has been stressed. This current step is the point in the process where the concept of market segmentation comes into play.

Adopting a More Tightly Defined Strategic Emphasis: An Illustrative Case

The author once worked with a company that developed computer-aided design and drafting (CAD) software for educational applications. This was in the early days when CAD technology was still in its infancy. The founder and CEO of the company came from a sales and marketing background and saw every design and drafting program in every community college and technical school in the country as his market. The words *market segmentation* were not in his vocabulary.

Some design and drafting programs specialize in teaching a particular drafting field such as mechanical, architectural, civil, structural, or electrical drafting. Others attempt to expose students to all of these fields. The CEO in this case decided that his company would be all things to all design and drafting programs. In other words, he planned to take the shotgun approach to marketing his company's products. This CEO thought his decision to focus on the education market and ignore the much larger private sector market consisting of engineering, architectural, construction, and manufacturing firms was enough market segmentation for any company. As things eventually turned out, he was wrong.

Developing CAD software for just one design and drafting field turned out to be a daunting task. It was much more time consuming, expensive, and knowledge intensive than the CEO had thought it would be. Developing software for all of the design and drafting fields commonly taught in community colleges and technical schools quickly proved to be an overwhelming challenge for an upstart company with limited resources. To add to the problems that began to quickly mount up for this company, the CEO—a marketing professional and eternal optimist—let his marketing campaign get ahead of his ability to produce the products. Before long, orders were pouring in for software that had been only partially developed.

Financial pressures forced the CEO to begin shipping partially developed, untested software to customers. The predictable results of this panic-based strategy weren't long in coming. In just a short time, customer complaints began to pour like rain, customer invoices went unpaid, opened software packages began to show up in the company's mail marked *return to sender*, and order cancellations began to outnumber new orders. Before long, the CEO found his company perched precariously on the edge of bankruptcy. To make matters worse, other CAD software companies that had heretofore focused on the private sector markets began to release less complex versions of their products at reduced prices that educational institutions could afford.

This is the point at which the author was asked for help. It was clear from the outset that the company would need to refine its market into a more tightly defined segment or segments and select a strategic emphasis that would allow it to be competitive. The CEO, faced with bankruptcy, accepted that his company could not be all things to all design and drafting programs. Unfortunately, by this time it was unclear whether the company could stay in business long enough to be anything to anybody. The prospects were grim. But the CEO was a fast and willing learner, and he was quite resourceful.

The CEO secured sufficient funds to stay in business for a while, and we worked together to adopt a strategic emphasis and a competitive strategy. As a competitive strategy, we chose to focus on architectural drafting because it appeared to be the market sector in which the company had the most expertise and the best chance of dominance. We chose a product-based strategic emphasis; that product would be CAD software for architectural drafting programs in community colleges and technical schools. Everything the company did from that point forward would be guided by this strategic emphasis.

With both an area of emphasis and a solid strategy in place, the company began to make some headway. Because the company was now tightly focused in an area where it had expertise, the CEO was able to guide his software experts through the development of a product designed especially for use in an instructional setting. Pedagogy was built into the software to make it, in essence, a self-teaching software package. This feature had immediate appeal because it allowed architectural design and drafting in-

| STRATEGIC PLANNING PROFILE | Competitive Strategy at Duke Power |

Duke Power is a major electric utility company that provides power to customers throughout the Carolinas. Changes in technology, legislation, and regulations have transformed the world of the electric utility company from the old days of virtual monopolies into an intensely competitive, global environment. Consequently, electric utilities such as Duke Power are now like other companies that must compete in a global business arena—they must adopt effective competitive strategies. Duke Power has done so. What follows are strategies Duke Power employs to stay competitive:

- Win and maintain the trust of all stakeholders, including employees, customers, regulators, and elected officials.
- Generate opportunities for sustainable sales growth.
- Satisfy customers by delivering valued products and services.
- Efficiently and effectively conduct all operations with superior safety, reliability, and responsibility.
- Effective management of cash, cost and capital, and win-win regulatory policy.

Source: Duke Power Web site (http://www.dukepower.com)

structors to focus on teaching their subject rather than having to conduct extensive computer training. The students liked the product because it accommodated self-paced learning; fast students could go fast and slower students could take more time.

As a result of adopting an appropriate strategic emphasis and an effective competitive strategy, the company began to make money before it ran out of funds. In fact, over time the company was successful enough that it was able to add new design and drafting fields to its product line. In every case, the software developed was self-instructional and self-paced in its design—a hugely popular feature with both instructors and students.

POTENTIAL AREAS OF STRATEGIC EMPHASIS

In the case just presented, the company adopted a product-based strategic emphasis, one of five commonly used areas of strategic emphasis that have potential for technology companies. These five areas of emphasis are:

1. Products or services
2. Competencies

3. Customers

4. Technology

5. Raw materials

These options are called areas of **strategic emphasis** because an organization going through the strategic planning process must choose one of them to emphasize. All of these areas are important to organizations that operate in a competitive business arena, and all of them should receive an appropriate amount of attention. But only one of these areas can be the *primary area of strategic emphasis.*

To understand this concept of selecting an area of primary strategic emphasis, consider the example of the professional football team. Blocking, running, tackling, passing, kicking, and pass defense are all-important and should receive an appropriate amount of attention from the team's managers and coaches. But one of these areas will be the team's primary area of emphasis, and the one chosen will define who the team is. Because offense is so important to professional football teams, the primary area of strategic emphasis is probably going to be either running or passing. This is why teams in the National Football League typically come to be known as either passing or running teams. They certainly do not ignore blocking, tackling, or pass defense, but they emphasize either running or passing.

Strategic Planning Tip

Key Point

These options are called areas of strategic emphasis because an organization going through the strategic planning process must choose one of them to emphasize. All of these areas are important to organizations that operate in a competitive business arena, and all of them should receive an appropriate amount of attention. But only one of these areas can be the *primary area of strategic emphasis.*

Product-Based Strategic Emphasis

Companies that define who they are by the product they produce have adopted a **product-based strategic emphasis.** In this case the term *product* also applies to services. The example of Microsoft Corporation illustrates this concept. Microsoft makes software. Everything the company is and does is defined and guided by this product—software. This does not mean that competencies, customers, technology, or raw materials can be ig-

Strategic Planning Tip

Key Point

Companies that define who they are by the product they produce have adopted a product-based strategic emphasis. In this case the term *product* also applies to services.

nored. Microsoft would not be a world-leading company if it ignored these critical areas. Rather, it means that of these various areas of emphasis, Microsoft places the highest level of emphasis on its product.

Because it has been so dominant in software markets globally, Microsoft has amassed the financial resources to branch out into virtually any field of business. However, in spite of this, the company sticks to one product—software—and focuses its efforts and energy on continually improving its software products. Companies that adopt a product-based strategic emphasis have made a conscious decision to make their product better than any other competitor in the marketplace can.

Organizations that adopt a product-based strategic emphasis should either have strong competencies in the following areas or plans to develop them:

1. New product development
2. Product design
3. Input, conversion, and output processes
4. Product quality
5. Continual improvement of process and product quality
6. Service quality relating to installation, maintenance, repair, and upgrading of products (as applicable)
7. Product marketing

Competency-Based Strategic Emphasis

Organizations that adopt a **competency-based strategic emphasis** are like gunslingers in the Old West. Theirs is a *have-gun-will-travel* approach, and their gun, so to speak, is a set of competencies. For the most part, they will produce any product or provide any service that fits within the realm of their competencies. An example from the past illustrates this point.

Westinghouse Corporation once had a plant that manufactured huge mechanical components for nuclear reactors. As a result, this particular plant had developed world-class competencies in the machining of stainless

Strategic Planning Tip

Key Point

Organizations that adopt a competency-based strategic emphasis are like gun-slingers in the Old West. Theirs is a *have-gun-will-travel* approach, and their gun, so to speak, is a set of competencies. For the most part, they will produce any product or provide any service that fits within the realm of their competencies.

steel components. When it came to machining high-quality metals such as stainless steel, this plant could broach, saw, turn, bore, drill, ream, mill, grind, cut, and weld them accurately to best-in-class tolerances. Then, almost overnight, the bottom dropped out of the nuclear reactor business world-wide. The nuclear reactor plant found itself in the peculiar position of having world-class competencies in the area of large-scale precision machining, but nothing to machine.

Executives at this Westinghouse plant knew they needed to find a way to stay in business while the parent organization decided what to do in the broader scheme of things. The obvious answer was to adopt a competency-based strategic emphasis, which the plant did. Rather than marketing a product, this plant began to market its world-class machining competencies. Any job that required precision machining of large parts made of stainless steel represented a potential contract.

Organizations that adopt a competency-based strategic emphasis should either have strong competencies in the following areas or plans to develop them:

1. Low costs relating to the competencies
2. Efficient input, conversion, and output processes (as applicable)
3. High-capacity capabilities
4. Substitute marketing
5. Flexibility relating to the application of the competencies

Strategic Planning Tip

Key Point

With the customer-based strategic emphasis, the organization focuses on a specific group of customers rather than a given product or service.

Customer-Based Strategic Emphasis

With the **customer-based strategic emphasis,** the organization focuses on a specific group of customers rather than a given product or service. This is a commonly used strategic emphasis. Companies that take this approach have the following philosophy of marketing: *We will provide every product you need if you fit into our specific customer group.*

Think, for example, of a school supply business. Such a company tries to provide every type of school supply needed by a customer base that consists of school children and their parents. Some companies apply the customer-based strategic emphasis to a wide variety of customer groups. For example, think of such retail outlets as Lowe's or the Home Depot. Each of these organizations provides a mind-boggling array of products, but their customer base is really tightly focused home improvers and homebuilders.

Organizations that adopt a customer-based strategic emphasis should either have strong competencies in the following areas or plans to develop them:

1. Brand loyalty
2. Customer relations
3. Ongoing customer research
4. Ongoing customer communication
5. Customer needs analysis capabilities
6. Ability to respond quickly to changing customer needs
7. Commitment to being a customer-focused organization

Strategic Planning Tip

Key Point

Companies with a true technology-based strategic emphasis are those that develop their business around a specific technology—typically though not necessarily one over which they have proprietary control.

Technology-Based Strategic Emphasis

The technology-based strategic emphasis is an often misunderstood concept. It does not automatically apply to organizations that are categorized as technology or high technology companies. Producing high technology products such as computers, electro-mechanical devices, or microprocessors does not equate to adopting a technology-based strategic emphasis. Most companies that produce high-tech products or have high-tech processes adopt the product-based strategic emphasis.

Companies with a true **technology-based strategic emphasis** are those that develop their business around a specific technology—typically though not necessarily one over which they have proprietary control. This strategic emphasis is similar to the competency-based emphasis explained earlier with one major exception. The latter involves building a business around a specific set of competencies that may or may not include a technological aspect. The former involves building a business around a specific technology, and the technology rather than a given set of competencies makes the company unique. The keys to success for companies that adopt the technology-based strategic emphasis are: (1) protecting the proprietary nature of the technology in question (when applicable), and (2) finding as many different applications of that technology as possible.

Organizations that adopt a technology-based strategic emphasis should either have strong competencies in the following areas or plans to develop them:

1. Proprietary rights to the technology in question (not essential, but a major advantage)
2. Applied research capabilities
3. Applications marketing
4. Flexibility in the applications of the technology
5. Ability to form and work well in strategic partnerships

Strategic Planning Tip

Key Point

The key to success in applying the raw material strategic emphasis is to identify a broad range of products that can be made from the raw material in question and then to be competitive in developing, manufacturing, and marketing those products.

Raw Material Strategic Emphasis

The **raw material strategic emphasis** is a commonly used, but necessarily limited approach. It has been the strategic emphasis for some of the world's best-known companies. Think of companies that use certain raw materials such as coal, trees, water, corn, soybeans, oil, cotton, tobacco, fish, and many other raw materials to make a variety of products associated with those raw materials. For example, a major oil company might extract oil from its own wells or purchase it from the Middle East and then use it to make any number of petroleum-based products such as gasoline, motor oil, plastic garbage bags, roofing shingles, or tar for road building. This is a raw

Strategic Planning Tip

Key Point

Before actually adopting a strategic emphasis for an organization, the strategic planning team should review some of the information it has collected and organized in earlier steps in the process as well as the competencies associated with each potential strategic emphasis. It is important to adopt a strategic emphasis that matches the organization's competencies or at least the competencies the organization plans to have.

material strategic emphasis. The key to success in applying the raw material strategic emphasis is to identify a broad range of products that can be made from the raw material in question and then to be competitive in developing, manufacturing, and marketing those products.

This strategic emphasis has enjoyed renewed interest in the past two decades as the recycling movement has, in effect, created a new array of raw materials. Old newspapers, worn-out tires, used plastic containers, and numerous other recyclable materials have become the raw materials for building new companies based on this strategic emphasis.

Organizations that adopt a strategic emphasis based on raw materials should either have strong competencies in the following areas or plans to develop them:

1. Access to and control of sufficient supplies of the raw materials
2. Securing the raw materials (e.g., extracting, cultivating, harvesting, mining)
3. Input, conversion, and output processes
4. Flexibility in creating various products from the raw materials

Adopting a Strategic Emphasis for an Organization

Before actually adopting a strategic emphasis for an organization, the strategic planning team should review some of the information it has collected and organized in earlier steps in the process as well as the competencies associated with each potential strategic emphasis. What are the organization's strengths and weaknesses, and how might they affect the choice of a strategic emphasis? What are the organization's core value-producing competencies, and how might they affect the choice of a strategic emphasis? What are the market segments that appear to have the most potential for the organization, and what type of strategic emphasis is most likely to help the

> ## Strategic Planning Tip
>
> ### Key Point
>
> After all factors have been considered, the strategic emphasis adopted should be the one that is most likely to allow the organization to dominate the market segments it defines as having the most potential. Remember, the purpose of adopting a strategic emphasis—in fact, the overall purpose of strategic planning—is to find a way for the organization to be a champion rather than just another player.

organization dominate those segments? What are the strengths and vulnerabilities of major competitors in the organization's most promising market segments, and how might these affect the choice of a strategic emphasis?

It is important to adopt a strategic emphasis that matches the organization's competencies or at least the competencies the organization plans to have. For example, an organization that is weak in such areas as new product development, product design, continual improvement of process, product quality, and product marketing should either avoid the product-based strategic emphasis or carefully plan how it will invest the resources required to develop the necessary competencies in these critical areas. This situation—needing certain competencies that are currently lacking—is one of the main reasons why the strategic planning team conducted an internal financial analysis earlier in the process. The results of that internal analysis come into play in such situations as this. Do we have the financial resources necessary to make the improvements that will be required by the adoption of a product-based strategic emphasis? Unless the answer to this question is in the affirmative, the strategic planning team must consider other areas of strategic emphasis.

An organization considering the customer-based strategic emphasis that has not achieved brand loyalty, is weak in the area of customer research, and has proven to be slow in responding to ever-changing customer needs should either consider another strategic emphasis or plan to develop the customer-related competencies it currently lacks.

These types of issues should be raised for any and all strategic emphasis options considered by an organization. It is critical that there be a good match between the strategic emphasis adopted and the competencies associated with the successful application of that strategic emphasis. A professional baseball team that has a weak pitching staff would not adopt pitching as an area of strategic emphasis when developing its plan to win the World Series—unless, of course, it planned to invest the resources necessary to develop a world-class pitching staff. The same concept applies to organizations that operate in a competitive business arena.

After all factors have been considered, the strategic emphasis adopted should be the one that is most likely to allow the organization to dominate the market segments it defines as having the most potential. Remember, the purpose of adopting a strategic emphasis—in fact, the overall purpose of strategic planning—is to find a way for the organization to be a champion rather than just another player.

SELECTING A COMPETITIVE STRATEGY

The concepts of competitive advantage and competitive strategy were explained in Chapter 1. At this point in the strategic planning process—selecting a competitive strategy—let's review these concepts to refresh your memory and understanding. There are two overall types of competitive strategies for achieving dominance in the market segments chosen—low cost and differentiation. Each of these strategies can be either focused or broad in scope. Consequently, there are actually four competitive strategies to consider at this point in the strategic planning process:

1. Low-cost/broad scope
2. Differentiation/broad scope
3. Low-cost/narrow scope (niche)
4. Differentiation/narrow scope (niche)

Low-Cost/Broad Scope Competitive Strategy

Organizations that adopt the low-cost/broad scope competitive strategy attempt to provide their products to the broadest possible customer base at the lowest possible cost. This strategy requires organizations to produce highly standardized products with a generic appeal to customers who are interested only or at least primarily in price. To succeed with this strategy, an organization must maintain efficient processes and low-cost relationships with its suppliers.

Differentiation/Broad Scope Competitive Strategy

Organizations that adopt the differentiation/broad scope competitive strategy attempt to differentiate themselves in the market segments chosen by providing products or services to a broad base of customers with attributes that are so valuable to the customers that they are willing to pay a higher price for them. This is how Federal Express originally differentiated itself from the U.S. Postal Service. Guaranteed overnight delivery, the ability to track packages, and in-office pickup were attributes many customers were willing to pay extra for and that the Postal Service did not provide at that time. The keys to success with this strategy are (1) identifying attributes that are different from those of competitors and are highly valued by customers, and (2) pricing the product or service high enough to cover the costs associated with providing them.

Low-Cost/Narrow Scope Competitive Strategy

This is the low-cost strategy applied to a limited segment or niche of a broader market. It could also be called the *low-cost niche strategy*. To succeed with this strategy, there must be a segment of a broader market made up of customers whose buying behavior differs in some way from that of the broader market. By focusing on this segment and tailoring its input, conversion, output, marketing, and service functions to meet the needs of the customers in it, an organization can be the low-cost provider and still achieve sustained profitability.

For example, in the field of computer-aided design and drafting (CAD) systems, the broad market includes architectural, construction, manufacturing, and several engineering applications such as civil, structural, mechanical, electrical, and aeronautical. A producer of CAD systems that chooses to hold down the price of its product by focusing narrowly and defining its market niche as just one of these applications is applying the low-cost/narrow scope competitive strategy.

Differentiation/Narrow Scope Competitive Strategy

This competitive strategy is a variation of the differentiation strategy. The principle difference is that with the narrow scope, the organization fo-

Strategic Planning Tip

Key Point

Once an organization decides who it is and where it is going, it will get to its intended destination by applying one and only one competitive strategy per market segment. It is important to select a competitive strategy that complements the strategic emphasis adopted and accommodates the organization's strengths, weaknesses, financial condition, and core value-producing competencies.

cuses on a more limited set of attributes and differentiates them even more than competitors that differentiate for a broader market. As a result, both the customer base and the list of product attributes are narrower than those of companies pursuing a differentiation strategy that is broader in scope.

What the Competitive Strategy Does

Before selecting one of the competitive strategies just reviewed, the strategic planning team should review one more time what a competitive strategy actually does for an organization. Recall that the strategic planning process answers three questions:

1. Who are we?
2. Where are we going?
3. How will we get there?

The strategic emphasis just adopted answers the first two questions. An organization's strategic emphasis will be product, competency, customer, technology, or raw material based. Once one of these options has been adopted, the organization and everything it does from that point forward should be defined by that option (Who are we?) and guided by it (Where are we going?). At this point, the strategic planning team has only one more question to answer: *How will we get there?*

Once an organization decides who it is and where it is going, it will get to its intended destination by applying one and only one competitive strategy per market segment. All of the competitive strategies just reviewed have merit, but the effectiveness of a given strategy will depend on circumstances. It is important to select a competitive strategy that complements the strategic emphasis adopted and accommodates the organization's strengths, weaknesses, financial condition, and core value-producing competencies.

Selecting the Optimum Competitive Strategy

When selecting the competitive strategy for a given market segment, the strategic planning team should answer the following questions:

1. Does the organization hope to grow in this market segment? If so, which competitive strategy is most likely to allow the organization to grow?

2. Does the organization hope to just hold its own in this market segment by achieving a growth rate that at least equals market growth in this segment? If so, which competitive strategy is most likely to allow the organization to hold its own?

3. Does the organization hope to reduce market share in this segment while still increasing profits (relinquish unprofitable portions of the market segment)? If so, which competitive strategy is most likely to allow the organization to reduce market share while increasing profits?

4. Does the organization hope to optimize this market segment by minimizing investments relating to it while maximizing profits? If so, which competitive strategy is most likely to allow the organization to optimize?

5. Does the organization, after having considered all relevant factors, hope to simply get out of this market segment, but without hurting our competitive position in other market segments? If so, which competitive strategy is most likely to allow the organization to withdraw without incurring market backlash?

Once these questions have been answered for a given market segment, an appropriate competitive strategy for that segment can be adopted. When selecting a competitive strategy, the strategic planning team should consider the organization's competencies and compare them with the competencies that will be needed to effectively implement the strategy. If competencies will have to be enhanced in order to effectively implement a given competitive strategy, the organization's financial condition must first be considered. Organizational strengths, weaknesses, and core value-producing competencies should be considered in relation to any competitive strategy considered for adoption, as should information about the strengths and vulnerabilities of the competition.

Summary

1. An organization's strategic emphasis tells who it is and where it's going. Executives need to know these things to lead the organization. Investors need to know these things to make investment decisions concerning the organization. Employees need to know these things to understand how to help the organization achieve its goals. Consequently, adopting a strategic emphasis is important.

2. There are five potential areas of strategic emphasis for technology-oriented organizations: a basis in product/service, competency, customer, technology, or raw material. While all of these areas might be important and should receive an appropriate amount of attention, only one of them should be emphasized.

3. Organizations that adopt a product-based strategic emphasis will need to have strong competencies in the following areas: new product development; product design; input, conversion, and output processes; product quality; continual improvement of process and product quality; service quality; and product marketing.

4. Organizations that adopt a competency-based strategic emphasis will need to have strong competencies in the following areas: low costs relating to the competencies; efficient input, conversion, and output processes (as applicable); high capacity relating to the competencies; strong substitute marketing; and flexibility relating to the application of the competencies.

5. Organizations that adopt a customer-based strategic emphasis will need to have strong competencies in the following areas: brand loyalty, customer relations, ongoing customer research, ongoing customer communication, customer needs analysis, response to changing customer needs, and customer focus.

6. Organizations that adopt a technology-based strategic emphasis will need to have strong competencies in the following areas: proprietary rights to the technology, applied research, applications marketing, flexibility in the application of the technology, and formation and operation of strategic partnerships.

7. Organizations that adopt a raw-materials based strategic emphasis will need to have strong competencies in the following areas: access to and control of a sufficient supply of the raw materials; capability of securing the raw materials; input, conversion, and output processes; and flexibility.

8. The question *How will we get there?* is answered by the competitive strategy adopted. There are four widely used competitive strategies: low-cost/broad scope, differentiation/broad scope, low-cost/narrow scope (niche), and differentiation/narrow scope (niche).

Key Terms and Concepts

Competency-based strategic emphasis

Competitive strategy

Customer-based strategic emphasis

Differentiation/broad scope competitive strategy

Differentiation/narrow scope competitive strategy

Low-cost/broad scope competitive strategy

Low-cost/narrow scope competitive strategy

Product-based strategic emphasis

Raw-materials based strategic emphasis

Strategic emphasis

Technology-based strategic emphasis

Review Questions

1. What three questions does the strategic planning process answer?
2. Why is it important for an organization to adopt an appropriate strategic emphasis?
3. Explain the concept of the product-based strategic emphasis.
4. What competencies should an organization that adopts a product-based strategic emphasis have?
5. Explain the concept of the competency-based strategic emphasis.
6. What competencies should an organization that adopts a competency-based strategic emphasis have?
7. Explain the concept of the customer-based strategic emphasis.
8. What competencies should an organization that adopts a customer-based strategic emphasis have?
9. Explain the concept of the technology-based strategic emphasis.
10. What competencies should an organization that adopts a technology-based strategic emphasis have?
11. Explain the concept of the raw materials based strategic emphasis.
12. What competencies should an organization that adopts a raw materials based strategic emphasis have?
13. Explain what an organization must do to succeed in applying the low-cost/broad scope competitive strategy.
14. What are the keys to success in applying the differentiation/broad scope competitive strategy?

15. What must an organization do to succeed in applying the low-cost/narrow scope competitive strategy?

16. List the questions the strategic planning team should answer when selecting the best competitive strategy for a given market segment.

SIMULATION CASES FOR DISCUSSION

The following simulation cases are provided to generate additional thought and discussion about the principles explained in this chapter. Readers are encouraged to consider how the situations presented in these cases might apply to them and to discuss the cases with others interested in strategic planning and execution.

CASE 7.1 | I Don't Understand Why We Adopt Just One Strategic Emphasis

"I'm telling you that our products, our customers, and the technologies that make up our conversion processes are all equally important," said Gail Parnell. Parnell is a member of the strategic planning team for Ortel Enterprises, Inc. (OEI).

"I agree with Gail," offered another member of the team. "I don't understand why we adopt just one strategic emphasis. Why not just adopt all three of these areas?"

Discussion Questions

1. Join this debate. Do you agree that OEI should adopt only one area of strategic emphasis for each of its market segments? Explain your reasoning.

2. What do you think the results of adopting all three areas of emphasis suggested by Gail Parnell might be for OEI?

CASE 7.2 | I Think We Should Adopt a Product-Based Strategic Emphasis

After discussing and debating the issue for more than an hour, the OEI strategic planning team decided to adopt just one area of strategic emphasis. Now the team must decide which area to choose.

"I think we should adopt a product-based strategic emphasis," offered Gail Parnell.

"You must be kidding," said another member of the team. "We are weak in the areas of new product development, product design, and product marketing. I think we should adopt a customer-based emphasis."

"We don't have strong enough brand loyalty or customer relations to adopt a customer-based emphasis," said Parnell in rebuttal.

Discussion Questions

1. Join this debate. Consider the competencies that are associated with the two options being discussed. Which option do you think would be best for the company, assuming it will have to enhance its existing competencies in either case?

2. If you were a member of OEI's strategic planning team, what information beyond that given in this case would you want to have before adopting a strategic emphasis?

CASE 7.3 | I Don't Understand How to Select a Competitive Strategy

"I understand why we selected the strategic emphasis we selected," said Gail Parnell. "But I don't understand how to select a competitive strategy. Can someone explain to me why we would choose a differentiation strategy instead of a low-cost strategy?"

Discussion Questions

1. Put yourself in this case. How would you explain the rationale for choosing the differentiation strategy rather than the low-cost strategy?

2. If the differentiation strategy is the most appropriate one for OEI, what can we assume about this company?

Endnote

1. As quoted in Boone, L. E., *Quotable Business*, 2nd ed. (Random House, Inc., New York: 1999), 31.

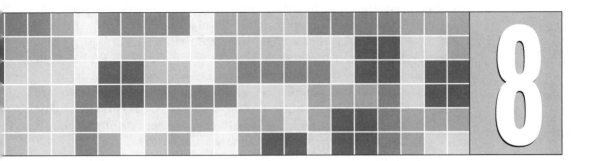

Write the Strategic Plan

It is a bad plan that admits of no modification.[1]

—Publilius Syrus

OBJECTIVES

- Explain how all of the work completed in the previous steps is brought together in this step to form a written strategic plan for the organization.

- Demonstrate how to develop the vision statement component of the strategic plan.

- Demonstrate how to develop the mission statement component of the strategic plan.

- Demonstrate how to develop the guiding principles component of the strategic plan.

- Demonstrate how to develop the strategic goals component of the strategic plan.

All of the steps that preceded this one were accomplished for the purpose of laying a solid foundation on which to build a strategic plan. In this step all of the foundational work accomplished thus far is pulled together into a concise, well-structured written document. Think of a strategic plan as an automobile. The chassis has been fabricated at one plant, the engine at another, and various other components at yet other plants. All of these various components have been shipped to the assembly plant and will now be pulled together to make a new car. This, in effect, is how a strategic plan is developed. In the current step, it's time to assemble the car.

Strategic Planning Tip

Key Point

For better or for worse, strategic planning is less standardized than many would like. Rather, it is one of those processes that is like hitting a baseball. There are a lot of different approaches, all of which will produce an excellent result if done well.

OVERVIEW OF THE STRATEGIC PLAN

Purchase any two guides to strategic planning, and you are sure to find differences in nomenclature and organization. Some strategic planning consultants call the *vision* what others call the *mission*. Some call the *guiding principles* what others call *core values*. Some conclude the strategic plan with the strategic goals, while others continue the plan to include specific projects, assignments, schedules, deadlines, and responsible parties. Technical professionals often express frustration over the fact that few planning consultants seem to take the same approach or use the same nomenclature.

This is a valid criticism. Technical professionals are accustomed to working with processes that are adjusted and refined until the optimum methods are identified and then standardized based on the optimum methods. For better or for worse, strategic planning is less standardized than many would like. Rather, it is one of those processes that is like hitting a baseball. There are a lot of different approaches, all of which will produce an excellent result if done well.

Planning consultants typically recommend the approach to strategic planning that has worked best in their experience, and so it is with the author of this text. The approach presented in this book is an approach the author has used with technology-oriented organizations for almost 30 years. During this time the process has been refined many times based on actual experience in guiding organizations through the strategic planning process. Consequently, the process described throughout this book and how it is all pulled together in this step has proven to be an effective way to hit the ball, so to speak.

Contents of a Completed Strategic Plan

Using the approach explained in this book, the contents of a completed strategic plan are as follows:

1. Vision statement
2. Mission statement
3. Guiding principles
4. Strategic goals

In developing each of these components, the strategic planning team must take into account various aspects of the information that was collected, organized, and analyzed in the previous steps. The team must also accommodate decisions that have already been made about market segments, the strategic emphasis for each market segment, and the competitive strategy for each segment.

What About Specific Projects, Assignments, Responsibilities, and Budgets?

The four components that make up the strategic plan—the vision, mission, guiding principles, and strategic goals—are the *strategic* aspects of the plan. Details that get more specific than these four components are, in the author's process, operational in nature. Operational elements of the plan are developed in the next step, *execution*.

Throughout this text, we have stressed that the strategic elements of the plan answer three questions: (1) Who are we? (2) Where are we going? (3) How will we get there? In developing the written plan, the answers to these questions are converted into the four elements of the strategic plan. In developing these elements, the strategic planning team will call upon all of

---

Strategic Planning Tip

Key Point

The mission statement, if properly written, will answer the question, Who are we? The vision statement, if properly written, will answer the question, Where are we going?

the information collected, organized, and analyzed in the previous steps as well as all decisions made in those steps.

How Does It All Tie Together?

The **mission statement**, if properly written, will answer the question, Who are we? The **vision statement**, if properly written, will answer the question, Where are we going? In answering these two questions, the mission and vision statements will encompass the strategic emphasis and competitive strategy developed in the previous step. The **guiding principles** will explain further who the organization is by summarizing what it believes. This summary of beliefs establishes the framework within which the organization will pursue its vision and mission. The **strategic goals**, if properly written, answer the question, How will we get there? In developing the strategic goals, the strategic planning team will consider the organization's strengths and weaknesses as well as those of the competition. It will also consider the opportunities and threats that were identified earlier.

What About the Details?

Specific projects, assignments, responsibilities, schedules, deadlines, and budgets answer a new question: What specifically do we need to do to get there? This is the fundamental question that guides the development of the *execution plan* developed in step 9 of the author's process.

It is important to separate the strategic and execution elements of the strategic plan. The former provides the *big picture* information for the organization, and the latter provides the specific *day-to-day details.* Although both the strategic and execution plans should be viewed as fluid documents that are open to change, the execution plan should change more often and with more regularity than the strategic plan.

For example, objectives and projects in the execution plan will be completed and replaced by other objectives and projects on an ongoing basis.

Strategic Planning Tip

Key Point

An organization's guiding force, the dream of what it wants to become, and its reason for being should be apparent in its vision. A vision is like a beacon in the distance toward which the organization is always moving.

Strategic objectives are of the nature that they are always being pursued, but are seldom completely achieved. For example, if an organization establishes a strategic goal of 5 percent annual growth in a given market segment, this goal continues every year. If the organization grows by 5 percent this year, it won't just stop. It will want to grow by another 5 percent next year.

DEVELOPING THE VISION STATEMENT

In developing the vision statement, the strategic planning team should remember that in addition to the statement's other characteristics, it should also answer the question, Where are we going?

An organization's guiding force, the dream of what it wants to become, and its reason for being should be apparent in its vision. A vision is like a beacon in the distance toward which the organization is always moving. Everything about the organization—its structure, policies, procedures, and allocation of resources—should support the realization of the vision. The vision conveys where the organization wants to go. Consequently, the organization's strategic emphasis should stand out as part of its vision.

In an organization with a clear vision, it is relatively easy to stay appropriately focused. If a policy does not support the vision, why have it? If a procedure does not support the vision, why adopt it? If an expenditure does not support the vision, why make it? If a position or even a department doesn't support the vision, why keep it? An organization's vision must be established and articulated by the strategic planning team and understood by all employees. The first step in articulating an organizational vision is writing it down. This is called the *vision statement*.

Writing the Vision Statement

A well-written vision statement, regardless of the type of organization, has the following characteristics.

- It is easily understood by all stakeholders.
- It is briefly stated, yet clear and comprehensive in meaning.
- It is challenging, yet attainable.
- It is lofty, yet tangible.
- It is capable of stirring excitement for all stakeholders.
- It is not concerned with numbers.
- It sets the tone for employees.
- It conveys the organization's strategic emphasis.

STRATEGIC PLANNING PROFILE   **Vision at Chevron**

When writing the strategic plan, no aspect of the process is more important than developing a powerful vision statement. The vision statement has to do so much. It must set the tone for the company, stir excitement in stakeholders, create unity among employees, and give all concerned a lofty yet tangible target for which to aim. It must serve as a steady, dependable beacon in the distance that the company can use to navigate through the inevitable troubled waters encountered on the sea of global competition. Chevron has adopted such a vision statement—a statement that meets all of these requirements. Chevron's vision statement is as follows:

... "to be the global energy company most admired for its people, partnerships, and performance."

This is a brief, simple, but powerful statement and an excellent example for other companies to emulate when developing their vision statements.

Source: Chevron's Web site (http://www.chevron.com)

From these characteristics it can be seen that crafting a worthwhile vision statement is a challenging undertaking. What follows are vision statements that satisfy the criteria set forth previously.

Vision Encompassing a Product-Based Strategic Emphasis

The Institute for Continual Improvement will be recognized by its customers as the provider of choice for organizational development products that are the best in the world.

Vision Encompassing a Customer-Based Strategic Emphasis

Pendleton Safety Company will be the leading supplier in the United States providing for all of the occupational safety needs of manufacturing firms.

Vision Encompassing a Competency-Based Strategic Emphasis

Horton Machining will be the leading provider of precision-machining operations to customers in the Midwest.

Vision Encompassing a Technology-Based Strategic Emphasis

Angleton Technologies will be the world leader in applying the X1–2 technology to a wide variety of applications.

Vision Encompassing a Raw-Materials Based Strategic Emphasis

Agri-Tech, Inc., will be the leading producer in the world of soybean-based products.

These vision statements illustrate the practical application of the criteria set forth earlier. Are these statements easily understood? Yes. Any stakeholder could read the vision statements and understand the dreams of the organizations they represent. Are they briefly stated, yet clear and comprehensive in meaning? Yes. Each of the statements consists of one sentence, but the sentence in each case clearly and comprehensively conveys the intended message. Are these vision statements challenging, yet attainable? Yes. Each vision presents its respective organization with the challenge of being the best in a clearly defined market. Being the best in the United States or in the world is a difficult challenge in any field, but it is an attainable challenge. It can be done. Are these visions lofty, yet tangible? Yes. Trying to be the best is a lofty challenge, but, still, it is achievable and therefore tangible. Pick a field, and some organization is going to be the best in that field. It could be this organization. Are these visions capable of stirring excitement among stakeholders? Yes. Trying to be the best in any endeavor is an exciting undertaking, the kind in which people want to take part.

Are these visions capable of creating unity of purpose? Yes. All five give stakeholders a common rallying cry. This is the type of response that happens when a sports team sets its sights on the championship. The players, coaches, fans, and management all rally around the vision, pulling together as one in an attempt to achieve it. Do these statements concern themselves with numbers? No. Numbers are left for later in the strategic planning process. Do these visions set the tone for employees? Yes. Clearly, the organizations in question are going somewhere, and employees are expected to do their part to ensure that the organizations get there expeditiously. Do these visions encompass the strategic emphasis for each respective organization? Yes. The strategic emphasis is clear in each vision.

DEVELOPING THE MISSION STATEMENT

We have just seen that the vision statement describes what an organization would like to be. It's a dream, but it's not "pie in the sky." The vision represents a dream that can come true. The organization's competitive strategy should be apparent in its mission statement. The mission takes the next step and describes *who* the organization is and *what* it does.

What follows are four different versions of the same mission statement. Each version of the statement tells who the company is and what it does, as any good mission statement must. The difference in each statement is found in the competitive strategy of the organization.

Mission Statement Encompassing a Low-Cost/Broad Scope Strategy

The Institute for Continual Improvement (ICI) is an organizational development company dedicated to helping technology companies continually improve their ability to compete in the global marketplace. To this end, ICI provides a wide range of low-cost organizational development products to a broad base of technology companies worldwide.

Mission Statement Encompassing a Low-Cost/Narrow Scope Strategy

The Institute for Continual Improvement (ICI) is an organizational development company dedicated to helping technology firms continually improve their ability to compete in the global marketplace. To this end, ICI provides a wide range of low-cost organizational development products to engineering firms in the Southeastern United States.

Mission Statement Encompassing a Differentiation/Broad Scope Strategy

The Institute for Continual Improvement (ICI) is an organizational development company dedicated to helping technology companies continually improve their ability to compete in the global marketplace. To this end, ICI provides high-quality, self-paced, computer-based training modules for in-house training applications in engineering firms worldwide.

Mission Statement Encompassing a Differentiation/Narrow Scope Strategy

The Institute for Continual Improvement (ICI) is an organizational development company dedicated to helping technology companies continually improve their ability to compete in the global marketplace. To this end, ICI provides high-quality, self-paced, computer-based training modules for in-house training applications in engineering firms located in Northwest Florida.

In the low-cost/broad scope version of the mission statement, the company develops a wide range of organizational products and makes them available at low prices to any kind of technology company worldwide. In the low-cost/narrow scope version of the mission statement, the company has narrowed the scope of its market to the Southeastern United States. In the differentiation/broad scope version of the mission statement, the company is differentiating its products on the basis of quality and such attributes as self-paced and computer-based learning. The scope of its market is broad—worldwide. The differentiation/narrow scope version of the mission statement is differentiating its products on the basis of quality and such attributes as self-paced and computer-based learning. However, the scope of the market in this case has been narrowed down to just Northwest Florida.

In developing the mission statement for any organization, one should apply the following rules of thumb:

- Describe the *who* and *what* of the organization, making sure the *who* component describes the organization and its market.
- Be brief, but comprehensive. Typically one paragraph should be sufficient to describe an organization's mission.
- Choose wording that is simple, easy to understand, and descriptive.
- Avoid *how* statements. How the mission will be accomplished is described in the strategic goals section of the strategic plan.
- Make sure the competitive strategy is clear in the statement.

DEVELOPING THE GUIDING PRINCIPLES

The guiding principles are an extension of the vision and mission in that they further answer the question, Who are we? Not all planning consultants include developing guiding principles in the planning process. But the author recommends them as an essential component of the strategic plan.

Strategic Planning Tip

Key Point

An organization's guiding principles establish the framework within which it will pursue its mission. Each guiding principle encompasses an important organizational value. Together, all of the guiding principles represent the organization's value system—the foundation of its culture.

An organization's guiding principles establish the framework within which it will pursue its mission. Each guiding principle encompasses an important organizational value. Together, all of the guiding principles represent the organization's value system—the foundation of its culture.

An organization's guiding principles establish the parameters within which it is free to pursue its mission. The following guiding principles are examples of the types that might be adopted by technology companies:

- XYZ Company will uphold the highest ethical standards in all of its operations.
- At XYZ Company, customer satisfaction is the highest priority.

- XYZ Company will make every effort to deliver the highest-quality products and services to its customers.
- At XYZ Company, all stakeholders (customers, suppliers, and employees) will be treated as partners.
- At XYZ Company, employee input will be actively sought, carefully considered, and strategically used.
- At XYZ Company, continual improvement of products, processes, and people will be the norm.
- XYZ Company will provide employees with a safe and healthy work environment that is conducive to consistent peak performance.
- XYZ Company will be a good corporate neighbor in all communities where its facilities are located.
- XYZ Company will take all appropriate steps to protect the environment.

From this list of guiding principles, the corporate values of XYZ Company can be discerned. This company places a high priority on ethics, customer satisfaction, quality, stakeholder partnerships, employee input, continual improvement, a safe and healthy work environment, consistent peak performance, corporate citizenship, and environmental protection.

With these values clearly stated as the organization's guiding principles, employees, supervisors, managers, and executives know the parameters within which they must operate. When ethical dilemmas arise, as they inevitably will in business, employees know they are expected to do the right thing. If safety or health hazards are identified in the workplace, eliminating them will be a top priority. If employees spend their own time participating in community activities, they know it will reflect positively in their performance appraisals, because XYZ Company values corporate citizenship.

DEVELOPING STRATEGIC GOALS

Broad strategic goals translate an organization's mission into measurable terms. They answer the third and final question in the strategic planning process: How will we get there? They represent actual targets at which the organization aims and will expend energy and resources trying to achieve. Broad goals are more specific than the mission, but they are still broad. Well-written strategic goals have the following characteristics. They:

- Are stated broadly enough that they don't have to be continually rewritten.
- Are stated specifically enough that they are measurable.
- Are each focused on a single issue or desired outcome.

Key Point

Broad strategic goals translate an organization's mission into measurable terms. They answer the third and final question in the strategic planning process: How will we get there? They represent actual targets at which the organization aims and will expend energy and resources trying to achieve.

- Are tied directly to the organization's mission.
- Are all in accordance with the organization's guiding principles.
- Clearly show what the organization wants to accomplish.

In addition to having these characteristics, strategic goals apply to the overall organization, not to individual departments within the organization. In developing its strategic goals, an organization should begin with its vision, mission, summary of organizational strengths and weaknesses, summary of opportunities and threats, results of the competitive analysis, and results of the market analysis. A point to keep in mind is that strategic goals should be written in such a way that their accomplishment would give the organization a sustainable competitive advantage in the marketplace.

Using the Work from Earlier Steps in Developing Strategic Goals

At this point in the process, the strategic planning team knows who the organization is and where it is going. The team has collected, organized, and analyzed a great deal of material that will now be helpful in deciding how the organization will get to its chosen destination. Information available to the strategic planning team includes summaries of strengths and weaknesses, projected market behavior, a summary of core competencies that create value (or the lack of them), strengths and vulnerabilities of the competition, and a summary of potential opportunities and threats. With this information available, the strategic planning team has much of the information it will need to develop strategic goals.

The following set of questions may prove helpful to the strategic planning team as it goes through the process of formulating strategic goals:

1. What strengths, if effectively exploited, will help the organization achieve its vision and mission?
2. What weaknesses, if not corrected, will inhibit the organization's ability to achieve its vision and mission?

3. What market behaviors looming in the future indicate that the organization should change a product, market, or some other aspect of the organization?

4. Are the organization's core value-producing competencies as strong as they will need to be to dominate the market sector in question?

5. Do any of our competitors have strengths that the organization should avoid or plan to circumvent?

6. Do any of our competitors have vulnerabilities the organization should exploit?

7. What new opportunities should the organization pursue?

8. What threats should the organization prepare to address?

9. What other things must the organization do to achieve its vision and mission that are not covered in questions 1–8?

The answers to these questions are used by the strategic planning team in developing strategic goals.

Converting Information into Strategic Goals

In writing strategic goals for an organization, the following steps should be observed:

1. *Assemble input.*   Circulate the vision and mission to all members of the strategic planning team. Ask all stakeholders to answer the following question: What do we have to accomplish as an organization to fulfill our vision and mission? Assemble all input received, summarize it, and prepare it for further review.

2. *Find the optimum input.*   Analyze the assembled input, at the same time judging how well individual's suggestions support the organization's vision and mission. Discard those suggestions that are too narrow or that do not support the vision and mission.

3. *Resolve difference.*   Proposed goals that remain on the list after step 2 should be discussed in greater depth in this step. Allow time for participants to resolve their differences concerning the goals.

4. *Select the final goals.*   Once participants have resolved their differences concerning the proposed goals, the list is finalized. In this stage the goals are rewritten and edited to ensure that they meet the criteria set forth earlier.

Cautions Concerning Strategic Goals

Before actually developing strategic goals for an organization, it is a good idea to become familiar with several applicable cautions. These cautions are as follows:

Strategic Planning Tip

Key Point

Tie all goals not just to the mission but also to the vision. All resources and efforts directed toward achieving the strategic goals should support the mission and the vision.

- Limit the number of goals—5 to 12 is usually a good range for most organizations. This is a rule of thumb, not an absolute. However, if an organization needs more than 12 objectives, it may be getting too specific at this point.

- Keep the language simple so that the goals are easily understood by all employees at all levels of the organization.

- Tie all goals not just to the mission but also to the vision. All resources and efforts directed toward achieving the strategic goals should support the mission and the vision.

- Make sure goals do not limit or restrict performance. This is best accomplished by avoiding numerical targets when writing strategic goals. Numbers, percentages, specific projects, time frames, and other such details come later when developing execution plans.

- Remember that achieving a strategic goal is a means to an end, not an end in itself (the vision is the end).

- Do not use strategic goals in the employee appraisal process. The only aspect of the overall strategic plan that might be used in the employee appraisal process is the specific-tactics component that comes in the next major step in the process—execution. This is because only the specific tactics in the strategic plan are assigned to the specific teams or individuals and given specific time frames within which they should be completed. Strategic goals, on the other hand, are everyone's responsibility. This is why execution is so important.

- Relate strategic goals to all employees. This means there should be goals covering the entire organization. Employees should be able to see that their work supports one or more of the broad objectives.

- Make strategic goals challenging but not impossible. Good goals will challenge an organization without being unrealistic.

Examples of Strategic Goals

What follows in this section are examples of strategic goals that were developed in response to the nine questions posed in the previous section. Each example shows how information from an earlier step in the process might be used to develop a strategic goal.

Strategic Goal Based on an Organizational Strength

Capitalize on our ability to quickly develop and introduce new products to preempt our competition in the ABC market segment.

Strategic Goal Based on an Organizational Weakness

Apply the Six Sigma concept to exponentially improve the performance of our conversion processes.

Strategic Goal Based on Perceived Market Behavior

Restructure our marketing function to make optimal use of Internet marketing and sales.

Strategic Goal Based on Weak Core Value-Producing Competencies

Improve our delivery system to the point that we can guarantee on-time delivery anywhere and every time.

Strategic Goal Based on a Competitor's Strength

Implement a marketing program that makes the higher purchase price of our product a plus when compared with the prices of our competitors.

Strategic Goal Based on a Competitor's Vulnerability

Implement a marketing program that exploits our product's higher reliability rating when compared with our principal competitor's products.

Strategic Goal Based on a Perceived Opportunity

Expand our software offerings by purchasing XYZ Software Development, Inc.

Strategic Goal Based on a Perceived Threat

Add the new XL-2 function to our product to preempt the planned entry of two new competitors into our principal market segment.

Strategic Goal Based on an Issue Not Raised by the Other Questions

Establish a companywide Effective Customer Service program to enhance relationships with all of our customers.

Summary

1. A complete strategic plan will have at least the following components: vision statement, mission statement, guiding principles (sometimes called core values), and strategic goals. Some planning consultants add specific projects, activities, and assignments to the strategic plan while others end the plan with the strategic goals and develop separate execution materials listing projects, activities, and assignments.

2. This book recommends ending the strategic plan with the strategic goals. Any thing beyond this level of planning is operational and execution-oriented. Execution-related planning materials are separate from the strategic plan in the model presented in this book. When writing the strategic plan, all of the work that has preceded this step comes into play and is pulled together in one comprehensive document.

3. The vision statement conveys where the organization wants to go. Consequently, the organization's strategic emphasis should stand out as part of the vision. The vision statement should have the following characteristics: easily understood by all stakeholders, briefly stated yet clear and comprehensive in meaning, challenging yet attainable, lofty yet tangible, capable of stirring excitement among stakeholders, capable of creating unity of purpose among all stakeholders, not concerned with numbers, sets the tone for employees, and conveys the organization's strategic emphasis.

4. The mission statement describes who the organization is and what it does. The organization's competitive strategy should be apparent in its mission statement. The following rules of thumb apply when developing a mission statement: (1) describe the who and what of the organization; (2) be brief but comprehensive (one paragraph should be sufficient); (3) choose wording that is simple, easy to understand, and descriptive; (4) avoid how statements; and (5) make sure the organization's competitive strategy is clear in the statement.

5. An organization's guiding principles establish the framework within which it will pursue its mission. Each guiding principle encompasses an important organizational value. Collectively, the guiding principles represent the organization's value system—the foundation of its organizational culture.

6. An organization's strategic goals translate its mission into measurable terms. Each goal represents a target at which the organization aims and will expend energy and resources trying to achieve. Well-written strategic goals have the following characteristics: (1) stated broadly enough that they don't have to be continually rewritten; (2) stated specifically enough that they are measurable; (3) focused on a single issue or desired outcome; (4) tied directly to the organization's mission; (5) in line with the organization's guiding principles; and (6) show what the organization wants to accomplish.

Key Terms and Concepts

Assemble input

Briefly stated, yet clear and
    comprehensive in meaning

Capable of creating unity of purpose

Capable of stirring excitement

Challenging, yet attainable

Easily understood by all
    stakeholders

Find the optimum input

Guiding principles

Lofty, yet tangible

Mission statement

Resolve differences

Select the final goals

Sets the tone

Strategic goals

Vision statement

Who and what of the organization

Review Questions

1. What are the four required components of a strategic plan?
2. Explain the rationale for separating the strategic plan and execution/implementation material for the plan.
3. What are the characteristics of a well-written vision statement?
4. Summarize the rules of thumb to follow when writing a mission statement for an organization.
5. What are guiding principles? What purpose do they serve?
6. List the characteristics of well-written strategic goals.
7. Summarize the steps used in actually writing strategic goals for an organization.
8. Explain the cautions to be observed concerning strategic goals.
9. Write a sample vision statement and a corresponding mission statement for a hypothetical company.
10. Write two sample guiding principles and two strategic goals for a hypothetical company.

SIMULATION CASES FOR DISCUSSION

The following simulation cases are provided to generate additional thought and discussion about the principles explained in this chapter. Readers are encouraged to consider how the situations presented in these cases might apply to them and to discuss the cases with others interested in strategic planning and execution.

CASE 8.1 I Want Specifics!

The debate had been going on for 20 minutes. The CEO of Mag-Tech, Inc. (MTI) wanted the strategic planning team to include specific projects, assignments, deadlines, and responsible parties in the plan. He summed up his point of view several minutes earlier when he said, "I want specifics!" The MTI planning consultant disagreed with the CEO. She wanted the strategic planning team to develop a plan that includes a vision statement, mission statement, guiding principles, and strategic goals. After this, she would guide the team through development of a separate execution plan covering specifics.

Discussion Questions

1. Who is right in this case—the CEO or the planning consultant? Explain your reasoning.
2. Can you think of a specific benefit in separating strategic and execution planning?

CASE 8.2 I Don't Like This Vision Statement

The planning consultant for MTI had been open and frank with the company's strategic planning team from the beginning. She didn't mince words. The vision statement the team proposed is as follows: *Mag-Tech, Inc., is in the business of manufacturing high-quality wheels for trucks and cars. We will grow in our market by 5 percent annually.* When the planning consultant read the proposed vision statement, she said, "I don't like this vision statement." MTI's CEO was surprised by her response and asked, "Why not? What's wrong with it?"

Discussion Questions

1. Step into the role of the planning consultant and answer the CEO's question in this case. What is wrong with the proposed vision statement?

2. Discuss a more appropriate wording of the vision statement for this hypothetical company.

CASE 8.3 This Is Not a Mission Statement

After the strategic planning team for MTI got its vision statement straightened out and properly worded, it developed a draft mission statement. The draft read as follows: *Mag-Tech, Inc., will be the leading producer of high-quality specialty wheels for trucks and cars in the United States.* The planning consultant's response upon reading the draft statement was, "This is not a mission statement." Once again the CEO, who played a major role in developing the draft, was surprised at her response. "What do you mean it's not a mission statement? What's wrong with it?" asked an obviously exasperated CEO.

Discussion Questions

1. Put yourself in the role of the planning consultant in this case and answer the CEO's question. What is wrong with the draft mission statement?
2. Discuss wording changes that would make this draft an appropriate mission statement.

CASE 8.4 I Want Some Guidance Before We Get Started

MTI's strategic planning team did a good job of developing a set of guiding principles for the company. The team was now ready to develop strategic goals. Before beginning the meeting, MTI's CEO took the planning consultant aside and said, "I want us to get this step right the first time. I want some guidance before we get started. Tell us how to go about developing strategic goals and give us a couple of examples."

Discussion Questions

1. Discuss the process you would recommend that the strategic planning team follow in developing strategic goals as if you were the planning consultant in this case.
2. Discuss the wording of a strategic goal that would give the CEO and other team members an example to follow.

Endnote

1. As quoted in Boone, L. E., *Quotable Business*, 2nd ed. (Random House, Inc., New York: 1999), 325.

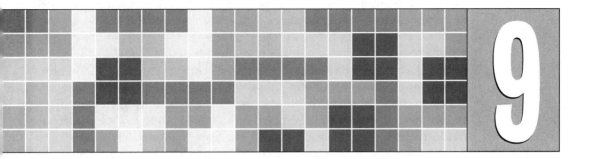

Execute the
Strategic Plan

Execution is not just something that does or doesn't get done. Execution is a specific set of behaviors and techniques that companies need to master in order to have competitive advantage.

—Larry Bossidy and Ram Charan,
Execution: The Discipline of Getting Things Done

OBJECTIVES

- Define the term *effective execution* as it applies in a strategic-planning setting.

- Explain why the execution phase of strategic planning so often breaks down.

- Demonstrate how to analyze a strategic plan for the purpose of anticipating roadblocks.

- Summarize effective execution strategies.

Every year, two teams from the National Football League meet in the Super Bowl. On game day, fans on every continent jam into their favorite sports bars and pack into each other's homes to cheer for their team, making the Super Bowl one of the most widely watched sporting events in the world. Knowing that an international spotlight is shining on them, both teams come to this annual event well prepared—each with its own game plan for winning. However, when the final whistle blows, only one team will emerge victorious and take home the coveted Lombardi Trophy. With rare exceptions, the winner of the Super Bowl is the team that does the best job of executing its game plan. The same can be said of companies operating in a competitive business arena.

The best strategic plan in the world is just a written summary of good ideas until it is effectively executed. It is one thing to plan, but it is quite another to execute. This chapter lays out tactics and techniques organizations can use to ensure that their change plans are effectively executed.

Strategic Planning Tip

Key Point

There is a saying that the devil is in the details. This saying certainly applies when executing a strategic plan. Whereas planning is about conceptualizing and envisioning with an eye to the future, executing is a nuts-and-bolts, hands-on enterprise that deals with the practical details of the here and now.

EFFECTIVE EXECUTION DEFINED

Effective execution as it relates to strategic plans is defined as follows:

> A rigorous and systematic process of initiating, following through on, and completing all tasks necessary to effectively implement the strategic plan. Execution involves such tactics and techniques as establishing expectations, assigning responsibility (delegation), establishing accountability, allocating resources, identifying and overcoming inhibitors, monitoring progress, flexibly adapting as necessary, and following up.

There is a saying that the devil is in the details. This saying certainly applies when executing a strategic plan. Whereas planning is about conceptualizing and envisioning with an eye to the future, executing is a nuts-and-bolts, hands-on enterprise that deals with the practical details of the here and now.

WHY STRATEGIC PLANS BREAK DOWN IN THE EXECUTION PHASE

This chapter opened with a Super Bowl analogy in which both teams came to the game with solid plans for winning the championship. Neither team will execute its game plan perfectly, but provided they both have good plans, the one that comes closest to an effective execution is the one most likely to win. Rarely is the Super Bowl won just because Team A had a better game plan than Team B. The teams that play in the Super Bowl know each other. The coaches on both teams are well versed in the comparative strengths and weaknesses of the teams and the individual players. Consequently, both coaching staffs are able to develop solid game plans that, if effectively executed, will ensure victory.

But developing a good game plan and executing it are two very different challenges. Plans are developed in the air-conditioned comfort of offices and conference rooms under conditions that are easily controlled. Plans are executed in the arena where your competitor is trying not only to execute its plan, but also to prevent you from executing yours. To illustrate this point, we return to the Super Bowl analogy.

Assume that a strategy in Team A's game plan is to run the ball from scrimmage 50 times during the game, and gain 250 rushing yards. This strategy is built into the game plan because Team A's coaches know from researching the opponent that the only teams that were able to beat Team B in the last three years were those that rushed for 250 yards or more. This is a well-founded strategy. However, it is much easier to put this strategy in the game plan than it will be to carry it out. During the game, every time Team A's quarterback calls a running play, 11 players on Team B will be doing their best to ensure its failure.

To gain 250 rushing yards in 50 carries, Team A will need to average five yards per carry. To accomplish this goal, every player on Team A must effectively execute his assignment on every rushing play—a difficult challenge under even ideal circumstances. But circumstances on the football field and in the business arena are rarely, if ever, ideal.

Strategic Planning Tip

Key Point

Developing a good game plan and executing it are two very different challenges. Plans are developed in the air-conditioned comfort of offices and conference rooms under conditions that are easily controlled. Plans are executed in the arena where your competitor is trying not only to execute its plan, but also to prevent you from executing yours.

> ### Strategic Planning Tip
>
> **Key Point**
>
> Another factor that often undermines the effective execution of strategic plans is what the author calls the *dirty hands syndrome*. This syndrome manifests itself in reluctance on the part of managers to step outside of the safe and comfortable intellectualism of planning and get their hands dirty sorting through the often-messy details of execution.

For example, consider Team B's game plan. Team B's coaches know that Team A typically depends more on its running game than on its passing game. Consequently, a strategy in Team B's game plan is to hold Team A to less than 100 rushing yards. This means that every time Team A and Team B line up across from each other, Team B's defense will be trying to undermine Team A's game plan by shutting down its running game.

This is exactly what happens to organizations trying to execute strategic plans. Company A decides to add a new product line. Company B and other competitors aren't going to simply sit back and let this happen. They will counter with their own new product or add features to an existing product to make it more marketable.

In addition, like football teams, organizations can be their own worst enemies when it comes to executing a plan. Football teams hurt themselves by fumbling the ball, missing blocks, throwing pass interceptions, and missing tackles. The corporate equivalents of fumbles, missed blocks, pass interceptions, and missed tackles can undermine the execution of strategic plans just as effectively as the efforts of competitors.

Dirty Hands Syndrome and Execution

Another factor that often undermines the effective execution of strategic plans is what the author calls the **dirty hands syndrome.** This syndrome manifests itself in reluctance on the part of managers to step outside of the safe and comfortable intellectualism of planning and get their hands dirty sorting through the often-messy details of execution.

These managers are like the architect who is comfortable sitting in an air-conditioned office engaged in the intellectual activities associated with designing a building, but is decidedly uncomfortable when it's time to put on a hard hat and visit the job site to check on the messy details of concrete being poured and steel beams being erected.

Larry Bossidy and Ram Charan explain as follows why even the best plans often break down in the execution phase:

Strategic Planning Tip

Key Point

Planning involves considering the possible, envisioning the ideal, and looking to the future. It is an appealing intellectual exercise full of promise and hope. Execution, on the other hand, involves digging into the details, dealing with reality, and focusing on the here and now. It is a practical, roll-up-your-sleeves-and-get-your-hands-dirty undertaking that is tempered by the practical realities of business and life. Small wonder then that so many managers find planning more appealing than execution.

. . . intelligent, articulate conceptualizers don't necessarily understand how to execute. Many don't realize what needs to be done to convert a vision into specific tasks, because their high-level thinking is too broad. They don't follow through and get things done; the details bore them. They don't crystallize thought or anticipate roadblocks. They don't know how to pick people for their organizations who can execute. Their lack of engagement deprives them of the sound judgment about people that comes only through practice.[1]

Planning involves considering the possible, envisioning the ideal, and looking to the future. It is an appealing intellectual exercise full of promise and hope. Execution, on the other hand, involves digging into the details, dealing with reality, and focusing on the here and now. It is a practical, roll-up-your-sleeves-and-get-your-hands-dirty undertaking that is tempered by the practical realities of business and life. Small wonder then that so many managers find planning more appealing than execution.

To grasp the difference between planning and executing, ask yourself the following questions:

1. Which would you rather do—plan a family budget or live within one?
2. Which do you find more appealing—planning a diet or sticking to one?
3. Which do you find easier to do—planning an exercise program or staying with one?

CONDUCT A ROADBLOCK ANALYSIS OF THE STRATEGIC PLAN

The best question the strategic planning team can ask before beginning the implementation phase is: *Where are the roadblocks to effective execution?* Any person given responsibility for any activity that is part of the plan should ask the same question. No enterprise that involves people, processes, money, and markets is going to simply unfold according to plan with no problems. There will always be roadblocks. One key to effective execution is accepting this fact and doing what is necessary to anticipate the roadblocks. If roadblocks are anticipated, contingencies can be planned for getting around them.

Roadblock Analysis: An Example

What follows are excerpts from the strategic plan for a hypothetical technology company:

- *Vision* Fire-Tech, Inc., will be the most customer-focused, customer-driven, customer-friendly company in the fire-suppression systems market in the United States.
- *Strategic Goals:*
  1. Establish high expectations for excellent customer service company-wide.
  2. Identify what customers value most about our systems.
  3. Benchmark our processes with special attention given to how they affect customer service.
  4. Compare actual performance against benchmarks and improve our processes continually.
  5. Provide customer service training companywide.
  6. Reclaim lost/dissatisfied customers.
  7. Establish effective, frequent, ongoing communication with customers.
  8. Establish a customer-oriented culture.

Each strategic goal in the plan is broken down into a list of implementation activities and projects. These projects become the focus of the roadblock analysis. The strategic planning team solicits input from personnel with expertise in the areas in question. For example, the implementation activities that correspond to specific strategy 1—*Establish high expectations for excellent customer service companywide*—are as follows:

1. Rewrite all job descriptions to include customer service.
2. Revise all performance appraisal forms to include customer service criteria.
3. Revise reward and recognition systems to include consideration of customer service performance.
4. Communicate expectations to all personnel companywide.

To analyze this set of projects for possible roadblocks, the strategic planning team will solicit input from several different groups: (1) the CEO and executive managers will be asked for input relating to revising the company's reward and recognition system and communicating expectations; (2) human resources personnel will be asked for input concerning revisions to job descriptions, performance appraisal forms, and the company's reward/recognition system; and (3) supervisors will join with the human resources personnel in providing input concerning revisions to performance appraisal forms.

The actual roadblock analysis consists of answering the following types of questions:

1. What factors might inhibit our efforts to complete this project fully, correctly, and on time?
2. Do we have the resources to complete this project fully, correctly, and on time?
3. Do we have the expertise to complete this project fully, correctly, and on time?
4. Are there existing habits, procedures, or informal ways of doing things that will inhibit our carrying out this project?

What follows is a roadblock that might have been discovered when Fire-Tech, Inc.'s strategic planning team conducted its analysis of the four activities tied to the first strategic goal in its plan:

- Neither the senior executives nor the human resources department personnel at Fire-Tech felt competent to revise the reward and recognition system without help. A solution to this roadblock would be to hire a compensation consultant to assist in the implementation of this project.

STRATEGIC PLANNING PROFILE   Effective Execution at Ferro Corporation

A strategic plan will not produce the desired benefits unless it is effectively executed. To effectively execute its strategic plan, a company must be as committed to the implementation phase as it is to the planning phase. This means that the CEO and executive leadership of the company must demonstrate their commitment to effective execution of the strategic plan. One way to show commitment to execution of the plan is to make its effective implementation one of the competitive strategies adopted by the company. Ferro Corporation does just this. One of the company's competitive strategies is an implementation strategy:

Implement the Leadership Agenda, a way of managing Ferro's portfolio of businesses to maximize profitable revenue growth, using more mature businesses to fund high-growth opportunities that represent Ferro's future.

Source: Ferro Corporation's Web site (http://www.ferro.com)

The analysis conducted by Fire-Tech's strategic planning team in this example identified a roadblock to the successful implementation of its first strategic goal. Had this problem not been identified and measures not taken to solve it, implementation of the company's strategic plan would have gotten bogged down at the outset. An analysis of the activities tied to the other strategic goals in the company's plan would, no doubt, reveal additional potential roadblocks. The roadblocks then become part of the next step—developing action/assignment sheets.

STRATEGIES FOR EFFECTIVE EXECUTION OF THE STRATEGIC PLAN

Everything done up to this point now culminates in the execution of the strategic plan. Execution is the most critical phase in the overall strategic planning process. According to Robert S. Kaplan and David P. Norton, a study of managers concluded that the ability to effectively execute strategy is more important than the strategy itself. [2] *Fortune* magazine reported that ineffective execution of strategy is a leading cause of CEO failure. [3] Clearly, executing the strategic plan is a critical challenge. The following strategies can help strategic planning teams improve their chances of having a successful implementation.

Develop Action/Assignment Sheets

The strategic plan contains goals that are written in broad, visionary terms. The organization has certain resources available for pursuing these goals. For example, the company has people, processes, and technologies. But even with all of these resources, there is still an essential ingredient missing. This missing ingredient is a device for transforming the strategic goals into specific activities, projects, and tasks that can be assigned to responsible personnel and placed within a timeframe.

The concept of the transformational device is best illustrated by an analogy. Assume you are the CEO of a successful technology company that has an opportunity to grow. Consequently, you adopt the following strategic goal:

> Construct a new corporate facility on a 50-acre tract of land outside
> of town.

You have identified a contractor who can do the work. This contractor has all of the resources needed to construct the facility (e.g., raw materials, people, technologies, processes). However, there is one critical ingredient missing—an ingredient without which this goal cannot be executed. The missing ingredient is a comprehensive set of blueprints.

The goal is written in such broad terms that the contractor has insufficient information to execute it. The contractor needs the goal to be trans-

---

Strategic Planning Tip

Key Point

People who will be responsible for executing parts of the strategic plan need an *execution blueprint* that transforms their responsibilities into operational terms.

---

formed into operational terms. The device for doing this is a comprehensive set of blueprints containing the architectural and engineering plans. The plans give the contractor the specific, detailed information he needs to construct the facility.

Like the contractor in this analogy, people who will be responsible for executing parts of the strategic plan need an **execution blueprint** that transforms their responsibilities into operational terms. These blueprints are not drawings, of course, but they do have all the characteristics of architectural and engineering plans in that they are detailed and specific. The author calls these execution blueprints **action/assignment sheets.**

Sample Action/Assignment Sheet

Action/assignment sheets contain all of the specific activities that must be completed to achieve a given strategic goal, the person or group to whom the activities are assigned, and projected completion dates. Figure 9.1 is an example of an assignment sheet that was developed for a hypothetical company.

The action/assignment sheet in Figure 9.1 was developed for the purpose of implementing one strategic goal from a large strategic plan. That goal, as shown in Figure 9.1, is as follows:

Establish and maintain a world-class workforce at all levels of the organization.

The strategic planning team determined that five specific tasks would have to be completed to set the organization on a path to accomplish this goal. Each of the tasks is assigned to a specific individual or unit that bears primary responsibility for implementation. The individual or unit in question is given a deadline for completing the assigned task.

Having a detailed action/assignment sheet for each strategic goal in the plan simplifies the process of execution by creating responsibility and accountability. An action/assignment sheet similar to the example in Figure 9.1 should be developed for all strategic goals contained in an organization's strategic plan.

Strategic Goal: Establish and maintain a world-class workforce at all levels of the organization.		
Activities	**Responsible Individual/Unit**	**Time frame/Deadline**
1. Arrange TQ training for all executive managers.	CEO	Completed by January 15
2. Arrange teamwork training for all executive managers.	CEO	Completed by January 30
3. Give all employees training in the use of problem-solving/ quality tools.	Department Managers	January 15–February 20
4. Give all employees training in continual improvement methods.	Department Managers	March 15–May 15
5. Establish a company supported off-duty education program for all employees.	Human Resources Department	In place by February 28

FIGURE 9.1   Action/assignment sheet.

Allocate a Budget for Each Action/Assignment Sheet

In the early stages of the planning process, the strategic planning team conducted an internal financial assessment. From that point on, financial *reality checks* have been a part of every step in the planning process. This is why it is so important for the organization's chief financial officer or someone else with financial expertise to be a member of the strategic planning team. These ongoing financial reality checks keep the organization from developing hundred-dollar plans on a ten-dollar budget.

Since every action taken by an organization has a cost associated with it, the strategic planning team should, with assistance of financial personnel, project the cost for each individual task on every action/assignment sheet developed. The funds necessary to accomplish the tasks in question are then allocated to those tasks. The individual or unit responsible for executing the task is also responsible for managing the funds allocated to it.

Figure 9.2. is an example of a budget allocation summary for funds allocated to the tasks listed in the action/assignment sheet in Figure 9.1. When the responsible individuals are given their assigned tasks, they are also given a budget allocation summary such as the one in Figure 9.2. Their job then becomes to execute the task within the budget allocated.

Strategic Goal: Budget Allocation Summary	
Establish and maintain a world-class workforce at all levels of the organization.	
Task	**Budget Allocation**
1. TQ training for executive managers	$10,000
2. Teamwork training for executive managers	$10,000
3. Problem-solving training	$50,000
4. Continual improvement training	$50,000
5. Off-duty education program	$35,000*
*$5,000 in start-up costs and $30,000 in annually recurring costs.	

FIGURE 9.2   Responsible individuals/units must know how much has been allocated for their respective tasks.

Establish and Maintain Momentum by Reinforcing Progress

Executing a strategic plan is like rolling a ball uphill. It's tough going at first, but if you can keep pushing until you crest the hill, gravity will take over and it's all downhill from that point on. When implementing strategic plans, the role of gravity is played by momentum—the tendency of an organization to keep going in a given direction once given a good start in that direction.

Once a sufficient number of action/assignment sheets have been developed and are being worked, execution of the strategic plan will begin to pick up momentum. If the organization can maintain this momentum, the execution ball will make it over the crest of the hill and begin the easier downward journey. To establish and maintain this level of momentum, use the following strategies:

1. *Model positive execution behaviors.*   While personnel at various levels in the organization are working on their assignments, it is important that they see executives, managers, supervisors, and members of the organization's leadership team working on theirs. Employees will take their cues from those above them in the organization. If they see executives, managers, and supervisors putting in the extra effort and time needed to carry out assignments, they will be more inclined to do so themselves.

2. *Talk about execution.*   Those individuals with responsibility for any aspect of an action/assignment sheet should talk up execution. Any

time employees are gathered together, the CEO and other executives and managers should talk about the progress of execution. It should be obvious from the comments of management personnel that executing the strategic plan is a high priority. By talking about the execution process, leaders keep stakeholders focused on it.

3. *Monitor progress.* Action/assignment sheets make it easy for those responsible to monitor the progress of execution on a daily basis. It is important to do so. This strategy can be combined with the previous strategy—talk about execution. Those personnel with specific responsibilities should keep copies of their action/assignment sheets close at hand. They should talk to every person with related tasks on their action/assignment sheet(s) every day and not just to ask about progress, but also to see actual evidence of progress. Even the most thoroughly thought-out assignments will overlook something. Even the most determined personnel will run into problems. By monitoring progress daily, management personnel can help those with assignments get around the execution problems that inevitably crop up so that work doesn't stop while the person with the assignment in question wonders what to do.

4. *Reinforce results by celebrating.* One of the best ways to maintain momentum is to reinforce progress by celebrating it. This can be done in a number of different ways. The quickest and one of the most effective is for executives to give public "attaboys" to personnel who complete their assignments. Letters of congratulations from the CEO, reports in company newsletters, small group meetings in which personnel are recognized for completing assignments, periodic written progress reports circulated among all employees recognizing the good work of personnel and teams that complete assignments, and companywide luncheons in which people and progress are recognized are just a few of the many ways progress can be reinforced through celebration.

Strategic Planning Tip

Key Point

By monitoring progress daily, management personnel can help those with assignments get around the execution problems that inevitably crop up so that work doesn't stop while the person with the assignment in question wonders what to do.

Summary

1. Effective execution as it relates to strategic plans is defined as a rigorous and systematic process of initiating, following through on, and completing all tasks necessary to the effective implementation of the plan. Execution involves such tasks and techniques as establishing expectations, assigning responsibility (delegation), establishing accountability, allocating resources, identifying and overcoming inhibitors, monitoring progress, flexibly adapting as necessary, and following through.

2. Strategic plans often break down in the execution phase for a variety of reasons. One of these reasons is that plans are developed in the air-conditioned comfort of offices and conference rooms, but they are executed in fluid circumstances in which the everyday realities of the marketplace, including the efforts of competitors, work to thwart the best-laid plans. Another factor that works against execution is the dirty hands syndrome. This syndrome manifests itself in reluctance on the part of executives and managers to step outside the safe and comfortable intellectualism of planning and get their hands dirty sorting through the often-messy details of execution.

3. The best question the strategic planning team can ask before beginning the execution of a strategic plan is this: Where are the roadblocks to effective execution? There will always be unforeseen circumstances and unanticipated problems. A roadblock analysis involves examining the plan thoroughly and objectively and asking the following types of questions for all tasks set forth in it: (1) What factors might inhibit our efforts to complete this project fully, correctly, and on time? (2) Do we have the resources to complete this project fully, correctly, and on time? (3) Do we have the expertise to complete this project fully, correctly, and on time? (4) Are there existing habits, procedures, or informal ways of doing things that will inhibit our carrying out this project?

4. To ensure that the projects and tasks in the strategic plan are executed fully, correctly, and on time, it is important to develop execution blueprints, called *action/assignment sheets*, and to reinforce progress as a way to establish and maintain momentum. Action/assignment sheets are developed for every individual strategic goal in the plan. An action/assignment sheet contains the actual assignment, the projected amount of time the assignment will take, the person responsible for the assignment, and beginning and projected completion dates. Funds allocated for completion of individual tasks are recorded in budget allocation summaries. Momentum is established and maintained when responsible management personnel model positive execution behaviors, talk about execution at every opportunity, monitor progress daily, and celebrate progress as it is made.

Key Terms and Concepts

Action/assignment sheets
Celebrate and reinforce results
Dirty hands syndrome
Effective execution
Execution blueprints

Model positive execution behaviors
Monitor progress
Roadblock analysis
Talk about execution

Review Questions

1. Define the term *effective execution* as it applies to strategic planning.
2. Explain why strategic plans so often break down in the execution phase.
3. What is the *dirty hands syndrome*, and how does it affect the execution of strategic plans?
4. What is a roadblock analysis?
5. What questions should be asked when conducting a roadblock analysis?
6. What is an action/assignment sheet, and what information should one contain?
7. What is meant by modeling positive execution behaviors?
8. Explain how talking about it promotes effective execution.
9. Why is it important to monitor progress closely when executing a strategic plan?
10. What is meant by celebrating results when executing a strategic plan, and why is it important to celebrate?

SIMULATION CASES FOR DISCUSSION

The following simulation cases are provided to generate additional thought and discussion about the principles explained in this chapter. Readers are encouraged to consider how the situations presented in these cases might apply to them and to discuss the cases with others interested in strategic planning and execution.

CASE 9.1 Executing the Plan Is Not Our Problem

"We developed the strategic plan. Let's give it to our department heads and let them get busy," suggested Joyce Duncan, a member of the strategic planning team at Safety Technologies, Inc.

"What about the details of execution?" asked Marvin Smith, also a member of the company's strategic planning team.

"That's not our problem," said Duncan. "Everybody who has anything to do with implementing our plan is a department head. Execution is not my job. I'm a manager. The department heads should be able to look at the strategic plan and determine what needs to be done in their areas."

Discussion Questions

1. What do you think of Joyce Duncan's approach to executing the company's strategic plan?
2. If you were taking part in this discussion, what advice would you give Duncan concerning executing the company's strategic plan?

CASE 9.2 That's Why Strategic Plans Don't Get Implemented

Marvin Smith and Joyce Duncan continue their discussion.

"That's why strategic plans don't get implemented, Joyce," said Marvin Smith. Smith and Duncan are engineering managers at Safety Technologies, Inc. Both serve on the company's strategic planning team.

"You can't just hand the various departments the strategic plan and tell them to get going," said Smith. "That would be like letting our engineers hand their rough design sketches to the manufacturing department and saying, 'Here, go make this'."

Discussion Questions

1. Do you agree with Marvin Smith in this case? Explain your reasoning.
2. What do you think will happen if the strategic planning team accepts the recommendation of Joyce Duncan to simply give a copy of the plan to all departments and tell them to get started?

CASE 9.3 We Need to Conduct a Roadblock Analysis

Marvin Smith and Joyce Duncan continue their discussion from the previous case. "Before we do anything else, we need to conduct a roadblock analysis of the strategic plan," said Smith.

"What are you talking about?" Duncan asked. "What is a roadblock analysis, and why do we need to do one?"

Discussion Questions

1. Do you agree with Marvin Smith that the strategic planning team needs to conduct a roadblock analysis as its next step?
2. How would you answer the last question posed by Joyce Duncan?

CASE 9.4 How Should We Go About Executing Our Plan?

Marvin Smith and Joyce Duncan continue their discussion from the previous case. "Alright, Marvin, we have completed the roadblock analysis you proposed, and I have to admit it was a good idea. Now can we get on with executing the strategic plan?" asked Duncan.

"We are almost ready," said Smith encouragingly.

"Almost! What else do we need to do?" asked Duncan.

Discussion Questions

1. Discuss how you would answer Joyce Duncan's last question in this case?
2. Discuss how you would recommend that the strategic planning team maintain momentum once the execution is underway.

Endnotes

1. Bossidy, L., and R. Charan, *Exectution: The Discipline of Getting Things Done* (Crown Business, New York: 2002), 36.
2. Kaplan, R. S., and D. P. Norton, *The Strategy Focused Organization* (Harvard Business School Press, Boston: 2001), 1.
3. Charan, R., and G. Colvin, "Why CEOs Fail," *Fortune*, June 21, 1999, cover story.

10

Understand How a Strategic Plan Comes Together from Start to Completion

A company intent on creating industry revolution has four tasks. First, the company must identify the unshakable beliefs that cut across the industry—the industry's conventions. Second, the company must search for discontinuities in technology, lifestyles, working habits, or geopolitics that might create opportunities to rewrite the industry's rules. Third, the company must achieve a deep understanding of its core competencies. Fourth, the company must use all of this knowledge to identify the revolutionary ideas, the unconventional strategic options that could be put to work in its strategic domain.[1]

—G. Hamel, "Strategy as Revolution"

OBJECTIVES

- Demonstrate an understanding of how a strategic plan is developed from start to completion.

- Demonstrate an understanding of how all of the individual elements of the strategic planning process relate to each other when developing a completed plan.

- Demonstrate how the individual elements of the strategic planning process are actually pulled together to create a comprehensive written strategic plan from start to completion.

- Demonstrate how to use a completed strategic plan to develop action/assignment sheets and corresponding budgets for the purpose of executing the plan.

The strategic planning process answers three questions for an organization: (1) Who are we? (2) Where are we going? (3) How will we get there? This chapter takes the reader through the entire strategic planning process undertaken by a hypothetical company. This case should help readers understand how the various elements of the strategic planning process fit together and how each respective component is actually developed.

DEVELOPING A STRATEGIC PLAN: DATA TECHNOLOGIES COMPANY

Data Technologies Company (DTC) started as a small minority-owned Department of Defense contractor. When the company was classified as an 8A firm, it became eligible for government set-asides. Set-asides under the 8A program are contracts awarded outside of the bidding process to minority-owned firms. The idea is to give such firms an opportunity to gain a foothold in business while learning how to compete without the set-asides. DTC entered this arena as a manufacturer of communications equipment for military aircraft. Its 8A status lasted for five years. During this time, DTC grew from a small shop in a garage to a company that employed almost 500 people.

When the company was just over a year from having to make the transition from 8A status to the competitive marketplace, its executive management team decided the company needed to create a strategic plan. For several years these executives had been so busy establishing the company and helping it grow that they had given little thought to what would happen to DTC once it graduated from the 8A program. Would the company be able to compete successfully in the open marketplace? Should it attempt to diversify into other markets? Should the executives simply sell the company and move on to other endeavors? These questions had begun to weigh heavily on the minds of DTC executives when the CEO suggested they hire a strategic planning consultant, go through the process, and see what transpires. DTC executives hired a planning consultant who led the company through the strategic planning process. The remainder of this section documents the process and its outcomes.

OVERVIEW OF THE PROCESS

The planning consultant worked with DTC's strategic planning team for two months on site at the company's headquarters. During this time the strategic planning team collected, organized, and analyzed information on DTC's financial condition, competitors, strengths, weaknesses, opportunities, threats, and core competencies that produce value. With all of this information summarized, the planning consultant arranged to take the team off site to write the strategic plan.

The consultant set up the strategic planning process in the conference center of a resort about 75 miles from DTC's facility. The idea, as he explained it to the strategic planning team, was to conduct the strategic planning process at a location that would guarantee both privacy and focus. No cellular telephones or visits to the office or home were allowed. In addition, family members were not included. The consultant explained that in order to come away with an acceptable draft of a strategic plan, the team would need three entire days of uninterrupted, fully focused work. The team would be given a morning and afternoon break each day to make telephone calls and check e-mail messages. Beyond that, their administrative assistants and family members knew how to reach them in case of an emergency.

The first hour of the first day was devoted to reviewing material and how it all fit together. The consultant explained all of the various components of the strategic plan (the vision, mission, guiding principles, and broad strategic goals). This three-day session would conclude with the development of a comprehensive draft of a strategic plan. In a follow-up session, the executives would develop the action-plan component. The action-plan session would involve a broader group that, in addition to the executive management team, would include other management and supervisory personnel. The session would be conducted by consultants after each member of the company's executive management team had solicited input from all these employees directly reporting to him or her.

REVIEW OF MATERIAL

Prior to beginning a review of material collected earlier in the process, the consultant placed several large flip charts in the room. The flip charts were labeled *strengths, weaknesses, opportunities, threats, core value-producing competencies,* and *competitor analysis.* Beginning with strengths, the consultant led participants through a brainstorming session. The purpose of the brainstorming was to use the material available to help the team adopt a strategic emphasis and competitive strategy.

Discussion was intense at times, and there were disagreements among participants. It took the consultant a while to convince participants to drop their defenses and to be open and frank without getting their feelings hurt or being territorial. Another dynamic was the differing perceptions of engineering, marketing, and finance personnel. Once the consultant worked through these and other issues that inevitably occur during strategic planning sessions, a cohesive group emerged and began to cooperate well as a team. The results of the review of material follow.

Strengths

Participants reviewed a summary of the following strengths: strong manufacturing capability approaching the Six Sigma level, solid business contacts in the Department of Defense industry, a proven track record of excellent performance in completing contracts on time, low turnover rate with regard to critical employees, comparatively low labor rates (most employees of DTC are retired military personnel who view their salary as a second income), and an up-to-date facility equipped with modern technology.

Weaknesses

Participants reviewed a summary of the following weaknesses: no experience outside of the Department of Defense market, no marketing component, no experience being the lead contractor on a major project (all of DTC's work up to this point had been subcontracted to it by larger Department of Defense contractors), no experience in the international marketplace, no design component (DTC had been a build-to-print operation up to this point), and no research and development component.

Opportunities

Participants reviewed a summary of the following opportunities: expansion into commercial aircraft markets, expansion into international commercial markets, expansion into foreign military markets (military aircraft of America's allies), and availability of a strong international marketing team that can work both commercial and military markets. (One of DTC's potential competitors in the commercial marketplace had just been purchased by a larger company, and its entire marketing team had been eliminated as part of the buyout.)

Threats

Participants reviewed a summary of the following threats: DTC's pending loss of its 8A status, potential cutbacks in the development of new military aircraft in the United States, a tight labor market that could inflate labor costs, and the potential for ever-increasing levels of competition from foreign and domestic sources.

Core Competencies That Produce Value

Participants reviewed a summary that showed DTC to have three critical core value-producing competencies: (1) the electronic assembly process, (2) mechanical packaging (consoles) of electronic components, and (3) ability to profitably produce small lot size production runs.

Competitors' Analysis

Participants reviewed a summary of the competitors' analysis completed early in the process that showed the following: none of DTC's competitors can accept small lot size production contracts. These competitors, all of which are larger than DTC, need large lot size production runs in order to be profitable.

DEVELOPING THE VISION

Before developing a vision for DTC, the company's strategic planning team had to decide if there would even be a DTC after graduation from the 8A program. Going ahead with the company would mean risking the investment of both the time and money of the company's executives. On the other hand, these executives could probably sell their shares in the company, walk away with a handsome profit, and find high-level positions with other firms in their respective fields or even help start another 8A company. After a lively discussion, the executives decided they had invested too much of their money and themselves in DTC to walk away from the company now. Consequently, they turned to the task of developing a post-8A vision for DTC.

The consultant led participants through a lengthy discussion that focused on the following questions: Should DTC stick with just domestic military markets or expand into the commercial marketplace as well? Should DTC consider pursuing contracts with foreign militaries? Should DTC pursue international commercial contracts? Should DTC stick with low-voltage power supplies as its principal product or diversify into other product lines? Should DTC add a design function or continue as just a build-to-print company? Should DTC add a research and development function to develop new product lines? In other words, the strategic planning team had to adopt a strategic emphasis and a competitive strategy.

Answering these questions was the most difficult part of the strategic planning process for DTC executives. How they answered these questions would determine everything else about the future of the company and, correspondingly, about their professional futures. After an intense discussion, participants decided that DTC would need to expand into both commercial and foreign markets while retaining its Department of Defense base. They also agreed that the company's expertise is in the production of electromechanical communication equipment for aircraft. Consequently, they ruled

<table>
<tr><td>STRATEGIC PLANNING PROFILE</td><td>Strategic Planning at Anheuser-Busch, Inc.</td></tr>
</table>

STRATEGIC PLANNING PROFILE — Strategic Planning at Anheuser-Busch, Inc.

Most people know Anheuser-Busch as a brewing company. But it is also the second largest producer in the United States of fresh-baked foods and the country's second largest theme park operator. With these diverse operations to oversee, strategic planning is important to the leadership team at Anheuser-Busch. The company does an excellent job in the area of strategic planning, as can be seen in the following excerpt from its plan:

Vision
Through all our products, services and relationships, we will add to life's enjoyment.

Mission
The mission of Anheuser-Busch is to:

- Be the world's beer company
- Enrich and entertain a global audience
- Deliver superior returns to our shareholders

Values
We believe in. . .

- Quality in everything we do
- Exceeding customer expectations
- Trust, respect and integrity in all of our relationships
- Continuous improvement, innovation and embracing change
- Teamwork and open, honest communication
- Each employee's responsibility for contributing to the company's success
- Creating a safe, productive and rewarding work environment
- Promoting the responsible consumption of our products
- Preserving and protecting the environment and supporting communities where we do business

Source: Jeffrey Abrahams, *The Mission Statement Book* (Ten Speed Press, Berkeley, CA: 1999), 71.

out adding a research and development function, but they did decide to expand into design. The rationale of DTC's strategic planning team was that the company would always be at the mercy of other, larger contractors unless it could design products in addition to just manufacturing them. The strategic planning team adopted a product-based strategic emphasis and a differentiation/broad competitive strategy.

With these questions answered, the consultant was able to lead participants through the process of developing a vision statement that would encompass the dreams of participants for DTC. The draft vision statement developed reads as follows:

> DTC will be an international leader in the production of electro-mechanical communication equipment for both military and commercial aircraft.

DEVELOPING THE MISSION

The strategic planning team found that developing a vision answered a lot of important questions. With the vision in place, developing a mission statement was not difficult; it was just a matter of following the criteria set forth by the consultant for well-written mission statements. Most of the discussion focused on wording as opposed to concepts. The draft mission statement the participants finally decided on reads as follows:

> DTC is a design and manufacturing firm dedicated to providing electro-mechanical communication products for the aircraft industry. To this end, DTC designs and manufactures communications equipment for military and commercial aircraft worldwide in small production runs.

DEVELOPING GUIDING PRINCIPLES

The consultant described *guiding principles* to participants as written statements that convey DTC's corporate values. He then encouraged the strategic planning team to mentally put the following sentence before the guiding principles to better understand what they represent: "While pursuing our vision and mission, we will apply the following guiding principles in everyday operations and in all decisions made."

The consultant asked participants to brainstorm important corporate values without concern, for the moment, about wording. Participants were encouraged to simply offer up value-laden terms (e.g., ethics, quality, customer satisfaction, etc.) that the consultant recorded on flip charts. Once the terms had all been selected, the consultant asked participants to select the six to ten most critical on the list. With the most important corporate values identified, participants worked with the consultant to develop more explicit wording for each one. The guiding principles developed are as follows:

1. *Ethics.*   All of DTC's employees and management personnel are expected to exemplify the highest ethical standards in doing their jobs.
2. *Customer delight.*   In dealing with customers, DTC will go beyond customer satisfaction to achieve customer delight.

3. *Continual improvement.*   Continually improving its products, processes, and people is a high priority for DTC.

4. *Quality.*   DTC is committed to delivering the highest-quality products on time, every time.

5. *Employee empowerment.*   DTC is committed to seeking, valuing, and using employee input and feedback.

6. *Partners.*   DTC is committed to treating its customers, suppliers, and employees as partners.

DEVELOPING STRATEGIC GOALS

The final component of the strategic planning session involved developing strategic goals. These goals had to represent broadly stated actions that, if accomplished, would move the company ever closer to the full realization of its corporate vision. Before beginning development of the goals, the consultant gave participants typed copies of the results of the SWOT analysis as well as the competitors' analysis and summary of core value-producing competencies. He explained to participants that the strategic goals developed should all satisfy one or more of the following criteria: (1) exploit one or more of the organization's strengths, (2) correct one or more of the organization's weaknesses, (3) take advantage of one or more of the opportunities available to the organization, (4) prepare against one or more of the threats facing the organization, (5) add or enhance a core value-producing competency, (6) exploit the vulnerabilities of a competitor, or (7) help DTC in some other way as it pursues its vision and mission. With this guidance, participants developed the following strategic goals for DTC:

1. Expand the company's business base to include both military and commercial markets in the United States and abroad.

2. Strengthen all functional units in the company in the area of commercial products and markets.

3. Expand the company's core value-producing competencies to include both design and manufacturing.

4. Fully achieve a Six Sigma quality level in the company's manufacturing function.

5. Develop and implement a supplier certification program to create a reliable group of dependable, high-quality supplier partners.

6. Establish a comprehensive training program to maximize the capabilities of all employees at all levels in the company.

With the strategic objectives established, the strategic portion of the plan was completed. The completed plan for DTC is shown in Figure 10.1.

Data Technologies Company
Strategic Plan

Vision

DTC will be an international leader in the production of electro-mechanical communication equipment for both military and commercial aircraft.

Mission

DTC is a design and manufacturing firm dedicated to providing electro-mechanical communications products for the aircraft industry. To this end, DTC designs and manufactures in small lot size production runs communications equipment for military and commercial aircraft worldwide.

Guiding Principles

1. *Ethics.* All of DTC's employees and management personnel are expected to exemplify the highest ethical standards in doing their jobs.

2. *Customer delight.* In dealing with customers, DTC will go beyond customer satisfaction to achieve customer delight.

3. *Continual improvement.* Continually improving its products, processes, and people is a high priority for DTC.

4. *Quality.* DTC is committed to delivering the highest-quality products on time, every time.

5. *Employee empowerment.* DTC is committed to seeking, valuing, and using employee input and feedback.

6. *Partners.* DTC is committed to treating its customers, suppliers, and employees as partners.

Strategic Goals

1. Expand the company's business base to include both military and commercial markets in the United States and abroad.

2. Strengthen all functional units in the company in the area of commercial products and markets.

3. Expand the company's core value-producing competencies to include both design and manufacturing.

4. Fully achieve a Six Sigma quality level in the company's manufacturing function.

5. Develop and implement a supplier certification program to create a reliable group of dependable, high-quality supplier partners.

6. Establish a comprehensive training program to maximize the capabilities of all employees at all levels in the company.

FIGURE 10.1   Completed strategic plan for Data Technologies Company.

PLANNING FOR AN EFFECTIVE EXECUTION

The off-site part of the strategic planning process arranged by DTC's planning consultant culminated, after three days of intense activity, in the completion of a comprehensive strategic plan. All that remained to be done at this point was to plan the activities and make the assignments that would ensure its effective execution. DTC's planning consultant and strategic planning team returned to the company's headquarters to develop action/assignment sheets for executing the strategic plan.

Developing Action/Assignment Sheets

The reason the execution planning phase of the process was conducted at DTC's headquarters was logistics. To develop action/assignment sheets that would direct the execution phase of the process, the strategic planning team and the planning consultant needed access to other personnel in the company who possessed specialized implementation-oriented knowledge. Rather than try to bring all of these various people to an off-site location—some of whom would be needed for ten minutes at a time and others who might be needed for hours—the strategic planning team and planning consultant agreed to complete their work on site.

Calling on the expertise of personnel from various functional units and departments, the strategic planning team developed an action/assignment sheet and corresponding budget for each strategic goal in the completed strategic plan. Figure 10.2 is an example of one of the action/assignment sheets developed. This action/assignment sheet was developed for the purpose of executing strategic goal 6 in DTC's strategic plan. *Establish a comprehensive training program to maximize the capabilities of all employees at all levels.*

Developing an action/assignment sheet for the execution of this strategic goal turned out to be more of a challenge than DTC's planning consultant had thought it would be. All members of the strategic planning team agreed that progress made toward achievement of this goal would be critical to the company's success. However, reaching consensus concerning how best to translate this goal into specific actions took a full day of often-heated debate.

Various members of the strategic planning team asked thought-provoking questions and raised interesting issues concerning how to break this strategic goal down into specific actions. The planning consultant had warned the members of the strategic planning team that "the devil is in the details" when developing action/assignment sheets. Several members of the strategic planning team commented that dealing with the larger and more conceptual issues of strategic planning seemed to be easier than dealing with the practical

ACTION/ASSIGNMENT SHEET

Strategic Goal:

Establish a comprehensive training program to

maximize the capabilities of all employees at all levels in the company.

Activities	Responsible Individual/Unit	Time frame/Deadline
1. Conduct a comprehensive training needs assessment at all levels in the company.	Human Resources/ Supervisors	Completed by January 20
2. Establish an in-house training academy for providing workshops, seminars, and short courses on site.	Human Resources/ Training Director	January 20 - March 20
3. Establish a partnership with a State University to offer accelerated business, management, supervision, and leadership courses on site.	Human Resources/ Training Director	January 20–April 1
4. Establish a company-supported off-duty education program with performance-based tuition reimbursement for approved courses of study.	Human Resources/ Training Director	January 20–February 20

FIGURE 10.2   One of DTC's action/assignment sheets.

details of execution planning. The planning consultant acknowledged that this is typically the case. Issues raised and questions asked by members of the strategic planning team included the following:

1. What types of training are needed at the various levels of the company?
2. What criteria should we use to determine whether training is appropriate and relevant?
3. Should the training take place on company time or after work hours or both?
4. Who should provide the training—DTC personnel, outside experts, or both?
5. Should the company pay the tuition costs of college courses taken during off-duty hours? If so, should the tuition be paid up front or on a reimbursement basis? If the reimbursement method is used, should

employees receive full reimbursement of tuition costs or a pro-rata percentage based on their grades?

After a lengthy discussion that involved DTC's Director of Human Resources who had been asked to join the strategic planning team as an in-house expert as the team dealt with strategic goal 6, several key decisions were finally reached. Those decisions are reflected in the action/assignment sheet in Figure 10.2. Before any type of training would be provided, the company's Human Resources Department—with help from first-line supervisors—would conduct a comprehensive, companywide training needs assessment. The needs assessment would attempt to answer the following question: What types of training are most critical to the various classifications/levels of employees at DTC in order for the company to be able to achieve its vision and mission?

The needs assessment led to the decision that some training would be provided in-house using talented DTC personnel as trainers while, at the same time, a partnership would be formed with a local university for providing higher levels of training. In addition, employees would be encouraged to take off-duty courses relating to their jobs or to prepare them for advancement in the company. Tuition would be reimbursed based on performance in classes taken, as demonstrated by the grades employees earned.

Each of the four actions relating to strategic goal 6 was assigned to a functional unit and/or individual along with a corresponding deadline for completion. The strategic planning team then sought the counsel of finance, human resources, and training personnel in allocating an appropriate budget for each action item. All of the strategic goals in DTC's new strategic plan were translated into action/assignment sheets in the same manner explained in this example.

This chapter illustrates all that is involved in developing a comprehensive strategic plan for a technology company from start to finish. Data were collected, organized, and analyzed; strengths, weaknesses, opportunities, and threats were identified; and the company's core competencies that allow it to produce value for customers were listed and described. Armed with this information, the strategic planning team developed a comprehensive strategic plan. The final plan included a vision, mission statement, guiding principles, and strategic goals.

Once the strategic plan was developed, action/assignment sheets were developed to ensure effective execution of the new strategic plan. The company in this example may be hypothetical, but the process is not. The process explained in this book and illustrated in this chapter is real, practical, and broadly applicable. It can be used by any technology company to plot a course for success.

Summary

1. A strategic plan answers the following questions for an organization: (1) Who are we? (2) Where are we going? (3) How will we get there? The process goes through many steps in answering these three questions. It is important, before beginning the development of a strategic plan, to understand how all of these steps fit together.

2. The internal assessment in which organizations identify strengths, weaknesses, financial condition, and core value-producing competencies produces information used as background when adopting a strategic emphasis and a competitive strategy. This information is also used when developing strategic objectives.

3. The external assessments and informed predictions organizations undertake to identify opportunities and threats as well as competitor strengths and vulnerabilities are used as background when adopting a strategic emphasis and competitive strategy and when developing strategic objectives.

4. The strategic plan encompasses four key elements: vision statement, mission statement, guiding principles, and strategic objectives. The organization's strategic emphasis and competitive strategy should be reflected in its mission and vision statements. The strategic goals should include any major weaknesses to be corrected, new markets to be entered, strengths to be exploited, core value-producing competencies to be added or strengthened, and any other broad actions that must be taken for the organization to establish and sustain competitive advantage in the marketplace.

Key Terms and Concepts

All of the key terms and concepts that appear in this chapter have been explained in detail in previous chapters. What is important in this chapter is to be able to present actual real-world examples of these key terms and concepts. Consequently, when reviewing the following terms and concepts, the reader is encouraged to think not so much of their definitions, which have already been adequately dealt with in earlier chapters, but of real-world examples that illustrate the term or concept.

Competitors' analysis
Core value-producing competencies
Developing action/assignment
    sheets
Developing guiding principles
Developing strategic goals

Developing the vision
Financial condition
Opportunities
Strengths
Threats
Weaknesses

Review Questions

1. Make an annotated flowchart that illustrates the strategic planning process from start to finish.
2. List all of the individual steps and elements within the steps of the strategic planning process and explain how they relate to each other.
3. Explain how all of the various steps of the strategic planning process are pulled together to create a comprehensive strategic plan.
4. Develop an action/assignment sheet for a hypothetical strategic goal.

SIMULATION CASES FOR DISCUSSION

The following simulation cases are provided to generate additional thought and discussion about the principles illustrated by example in this chapter. Readers are encouraged to consider how the situations presented in these cases might apply to them and to discuss the cases with others interested in strategic planning and execution.

CASE 10.1 Should We Go Off Site or Stay On Site?

The CEO of GFW, Inc., and her strategic planning team are discussing the logistical aspects of developing a strategic plan for their company. Some members of the strategic planning team think all of the work should be done in-house using the company's main conference room as the so-called "war room" for developing the strategic plan. Other members think the whole process should be conducted off site. After listening to both sides present their views, the CEO said, "Alright, let's decide. Should we go off site or stay on site?"

Discussion Questions

1. There are no hard and fast rules concerning on-site versus off-site locations for developing strategic plans. Which of these approaches do you think would work best and why?
2. DTC, the hypothetical organization in this chapter, used a combination of the two approaches. Internal assessments and informed predictions were completed on site. Development of the written plan took place off site. Then the execution planning took place on site. What do you think of this approach?

CASE 10.2 How Does All of This Come Together?

The GFW strategic planning team stood looking at what appeared to be a mountain of information. There were summaries of the company's financial condition, strengths, weaknesses, opportunities, threats, core value-producing competencies, strengths and vulnerabilities of competitors, options for selecting a strategic emphasis, and options for selecting a competitive strategy. Speaking both to herself and to the strategic planning team, GFW's CEO said: "I'm going to pull my hair out! Somebody tell me how all of this fits together."

Discussion Questions

1. Put yourself in the room with GFW's strategic planning team. What would you tell the CEO about how all of the information available to the strategic planning team fits together?

2. What would you tell the CEO and the team about how to convert all of the information it had collected, organized, and analyzed into a completed strategic plan?

CASE 10.3 How Are We Going to Make Sure This Plan Gets Executed?

GFW's CEO knew that just having a comprehensive strategic plan would not be enough to ensure that her company gained and sustained competitive advantage in the marketplace. She was aware of how important effective execution would be. However, she wasn't quite sure how to go from planning to execution. She put the matter before the company's strategic planning team. "How are we going to make sure this plan gets executed?" she asked.

Discussion Questions

1. Put yourself in the room in this case as a member of GFW's strategic planning team. How would you answer the CEO's question?

2. What problems might GFW's strategic planning team run into when trying to develop action/assignment sheets to guide the execution of its strategic plan?

Endnote

1. Hamel, G. "Strategy as Revolution," *Harvard Business Review* (July-August 1996), 80.

Index

Page numbers in *italics* indicate figures.